Praise for **HOW TO BE ALIVE**

"One of our most thoughtful role models is back with a deep, detailed, and generous Baedeker for how to live a better life."

—Seth Godin, author of *What to Do When It's Your Turn*

"*How to Be Alive* is rich in wisdom. Read it, reflect on it, and choose how you will live the rest of your life. It might turn out to be the most important book you have ever read."

—Peter Singer, author of *Animal Liberation* and *The Most Good You Can Do*

"Bold, loving, and wise . . . could be considered an essential piece of the map for living in the twenty-first century. Most inspiring is the way Colin Beavan revitalizes an ancient understanding—that our happiness and the happiness of others are interdependent."

—Sharon Salzberg, author of *Lovingkindness* and *Real Happiness*

"Colin Beavan's book will help you inflict untold goodness upon the world—all the while making you a happier person. Highly recommended for any human with a conscience."

—A. J. Jacobs, author of *The Year of Living Biblically* and *Drop Dead Healthy*

"For too long, being happy and doing good in the world have been presented as divergent life paths . . . Beavan convincingly upends this false choice. It turns out that doing good actually makes us happier and healthier. *How to Be Alive* is for anyone who wants both a better life and a healthier planet."

—Annie Leonard, executive director of Greenpeace USA and author of *The Story of Stuff*

"This book is for those who are brave enough to step away from fear, guilt, and self-centeredness and travel the hero's path. To dare to care about the happiness of others, including other animals, is to ensure your own happiness. In *How to Be Alive*, Colin Beavan reveals his secrets and shows us by his own living example that it is not so difficult to be a happy hero."

—Sharon Gannon, cofounder of Jivamukti Yoga

"At a time when the conventional definition of success is material wealth without limits, *How to Be Alive* offers a needed corrective. Colin Beavan starts local with his neighborhood and expands outward to the entire world, showing how we can all live better. When two-thirds of college students believe they'd rather make a killing than a difference, Beavan warns that such a course will kill their spirits and dampen their joy. When the market is flooded with self-help "happiness" books that are both vacuous and self-indulgent, Beavan makes clear that only by helping others can we truly help ourselves and live better. With dozens of imaginative ideas, he offers readers ways to heal themselves that can heal our planet and a divided humanity as well."

—John de Graaf, coauthor of *Affluenza*, president of Take Back Your Time, and cofounder of the Happiness Alliance

"Colin Beavan is a wonderful human and wonderful writer—and therein lies the genius of this book, where he circles in on answers to the unquiet so many of us have: how can I be real and be good and be right and be happy in a world unhinged from its moorings?"

—Vicki Robin, coauthor of *Your Money or Your Life* and author of *Blessing the Hands That Feed Us*

ALSO BY **COLIN BEAVAN**

No Impact Man:
The Adventures of a Guilty Liberal Who Attempts to Save the Planet, and the Discoveries He Makes About Himself and Our Way of Life in the Process

Operation Jedburgh:
D-Day and America's First Shadow War

Fingerprints:
The Origins of Crime Detection and the Murder Case That Launched Forensic Science

HOW TO BE ALIVE

A GUIDE TO THE KIND OF HAPPINESS **THAT HELPS THE WORLD**

COLIN BEAVAN

DEY ST.
AN IMPRINT OF
WILLIAM MORROW *PUBLISHERS*

ENVIRONMENTAL BENEFITS STATEMENT

HarperCollins Publishers LLC saved the following resources by printing the pages of this book on chlorine free paper made with 100% post-consumer waste.

TREES	WATER	ENERGY	SOLID WASTE	GREENHOUSE GASES
229	**107,126**	**103**	**7,171**	**19,752**
FULLY GROWN	GALLONS	MILLION BTUs	POUNDS	POUNDS

Environmental impact estimates were made using the Environmental Paper Network Paper Calculator 3.2. For more information visit www.papercalculator.org.

DEY ST.

This is a work of nonfiction. Some names have been changed in order to protect the integrity and/or anonymity of the individuals involved.

Printed on recycled paper.

FIRST EDITION

Designed by Paula Russell Szafranski

Library of Congress Cataloging-in-Publication Data has been applied for.

ISBN 978-0-06-223670-8

16 17 18 19 20 OV/RRD 10 9 8 7 6 5 4 3 2 1

For the little girl—rapidly becoming

young woman—who has taught me how to love.

I am so, so proud of you. My daughter,

Isabella Conlin Beavan

Likewise, helping oneself and helping others

Are like the two wings of a bird.

–ZEN MASTER WON HYO

Contents

Introduction

WHAT KIND OF LIFE DO YOU WANT?

The End of the World as We Know It and the Amazing Opportunities That Follow

Climate-change-induced superstorms that have killed many thousands and put tens of thousands out of their homes. An economy that seems permanently too weak and stingy to offer dependable jobs. Social and racial inequality filling the news and making us feel we are going backwards. A world political system that is too broken, deadlocked and corrupted by money to deal with any of it.

Everything seems so suddenly unstable.

We can all choose to live in fear about that. Or, we can look to a new set of aspirations and life choices that many people are already finding exciting and inspiring.

I don't mean the problems and the suffering are exciting, of course. I mean the opportunities are. Because as things shift, you (and I and everyone else) have the opportunity to make new choices and build new lives where you not only get to feel secure and enjoy your life but also get the chance to contribute to solutions that help other people, and feel like your life really means something in the process.

That's what this book is about—the quest for a joyous and meaningful life while living in a frightening, confusing world that needs our help. I know such a life is possible because, as I'll explain shortly, I've somewhat famously begun to experience it myself, and I've seen it in many thousands of others.

But first, listen:

It used to be that you had to follow a very specific path to get what you were supposed to want. Now, a college education no longer guarantees a corporate job, and a corporate job no longer guarantees health care. Even "securing" your retirement may seem like a pipe dream.

Meanwhile, a lot of us who do manage to "get ahead" can't stop feeling a sense of futility. We sit in our cubicles all day tapping at our keyboards while *not* using the talents we prize in ourselves. Not only do we not believe in the missions of the companies we work for, but we often find ourselves willfully ignoring the harm our employers cause. We can't escape the nagging feeling that our "success" comes at the expense of the rest of the world.

The standard approaches to life no longer lead where so many of us actually want to go—not when it comes to feeling as though we're helping with world problems and not when it comes to having secure and meaningful lives of our own.

Which leads us back to the exciting parts—the opportunities.

The first exciting part is this: since fitting into societal molds no longer pays off in the ways it traditionally did, we are freer to stop forcing ourselves into those molds. With fewer so-called rewards to supposedly miss out on, we have less to lose if we break away from the societal directions people traditionally follow and much to gain by experimenting with life choices that are actually truer to our values, our passions, and our world concerns.

The second exciting part is that in this time of numerous crises, how we choose to live can truly make a difference in the world. Back when the world was stable, it felt unchangeable. What impact could one person have? Now the world is a changing and fluid place.

Each of us is like a butterfly whose wing flaps could start a hurricane. A world with many ills needs many different kinds of doctors—many different kinds of people with many different kinds of talents and passions and personalities. We all have the chance to matter in an entirely new way.

The third exciting part is that we don't have to embark on this quest for the happy, impactful, values-based life alone, nor do we have to figure out the quest entirely for ourselves. It is true that our culture constantly nudges us toward the work-to-spend treadmill that so many of us want to escape. There is still no societal pathway to the authentic, meaningful, service- and passion-oriented life. But it is also true that there is already a growing national and international movement of people breaking away from those old broken paths and laying down new ones. They are questing for and finding new ways of making life choices—in careers, in what they do with their money, in their living situations and lifestyles, in how they eat and travel, and in all the many other ways we relate to each other.

As fate would have it, I became a well-known figure in this movement of seekers back in 2007, when I launched a one-year lifestyle experiment in environmental living. It was in many ways the climax of my lifelong quest to find a fulfilling, meaningful, happy life that helped others and was in line with my values. This year-long project became the subject of an autobiographical book and a documentary film, both titled *No Impact Man*.

The book has been translated into thirteen languages and is required reading on hundreds of college campuses. The Sundance-selected film has been screened in cinemas and broadcast by television networks around the world. Most important, this project, book, film, and the work that has evolved from my experience since have put me in touch with literally tens of thousands of seekers who are on the quest for a better way of life.

I've met and given talks to thousands of people who are creating their own new ways of relating to the society we live in. They are making lives that are better for themselves, better for their communities, and better for the world. You'll find some of their stories and much of their wisdom in the pages of this book.

The world needs entrepreneurs who use business as a tool for increasing happiness. It needs activists who speak with love instead of fear and anger. It needs gardeners and local farmers who care for the land. It needs a whole different kind of bankers and politicians who care more about communities than corporations. It also needs more musicians on the subway platforms and artists on the streets to bring us joy in these difficult times. The world needs so much. It needs all of us.

Which is not to say this is a job or career book, because it's not. Thinking that careers and jobs are the only way to security and meaning and helping the world is another of those standard life approaches we need to move away from. How we work matters, yes, but so do how we make friends, have families, think about possessions, run our homes, live in our communities, engage as citizens, have sex, relate to children, and on and on.

In this book, we are going to discuss the choices that will help us build great lives—not just great careers—based on who each of us really is, what we really care about, and what we most want to help with in the world. It is a book about asking the questions that will nudge you along a path to your own personal version of the Good Life. A life where your happiness and safety come not at the expense of the world but as the result of doing good for the world.

That idea, by the way, is what distinguishes this book from most other self-help books and why, as we will discuss later, I actually think of it less as a self-help book and more as an each-other help book. There is a multitude of books about the world's problems. And there is a multitude of books about how to try to extract a happy life from that world. What's missing are the books about fixing our lives in ways that fix the world and fixing the world in ways that fix our lives. That's what this book is about.

More About How I Came to Write This Book and How You Can Use It to Claim Your Own Version of a Cool, Fun, Meaningful Twenty-First-Century Life

On the door to the dressing room is a big star with my name on it, something I thought only happened in movies. Inside, I'm sitting on a couch. Now standing up. Now sitting. Now standing. I'm so nervous. I've already been on *Good Morning America* and many other shows, but the host of this show is one of the fastest wits on television. Going on camera with him is like throwing yourself under a rhetorical bulldozer—in front of a few million people.

I've already been to hair and makeup, where they found a way to eliminate the circles under my eyes. Now a woman with a clipboard arrives. She suddenly freezes as she listens to something in her headset. "Yes . . . Yes . . . I'm with him." She looks at me. "Come this way please, Mr. Beavan."

We walk down a hallway and a door opens to the studio. We pass what look like bleachers, where the studio audience is sitting. A comedian is warming them up and they are all laughing. I step up onto the stage and suddenly I am seated across a desk from Stephen Colbert of *The Colbert Report*.

This makes the first of what would ultimately be two appearances on the show. Apparently, despite Colbert's false conservative persona, he is a fan of *No Impact Man* and my work in lifestyle redesign.

I'm sitting up there onstage when, suddenly, the comedian who has been warming up Colbert's audience finishes and exits. The APPLAUSE sign is blinking over the audience. The little light above the camera lens has turned green. Colbert is introducing me. He jokes in his deadpan way, "By not poisoning the earth, he is poisoning our capitalist society."

God help me.

Before the *No Impact Man* project, I had tried so hard to get happiness in all the ways we are told will bring it: job searches, career "hacks," romantic relationships, following gurus, "attracting my vision," living in this city, living in that city, traveling, working full-time, working part-time, and so on.

As innovatively as I tried to approach all this, I still clung, without realizing it, to an old-fashioned set of societal directions. These directions, the *standard life approaches,* as I call them, all go along these lines: work like crazy to get some better thing, person, house, or job, which will then make you feel better in some way.

But for me—and probably you, since you are reading this book—it doesn't actually work. Many of us, though, have to hit some sort of bottom before we are willing to go against the cultural flow and try something new, which is what happened to me.

Back during the Iraq War, I happened to be writing my second history book, about a secret operation where Allied teams dropped behind enemy lines to work with the Resistance in Occupied France during World War II. While researching that book, I interviewed in depth more than seventy veterans of the operation and listened as they told their sometimes terrifying stories of what they had done and witnessed and what had been done to them. Some of them cried during their interviews. By the time I was done, having heard so much about the terrible things people can do to each other under desperate circumstances, I felt as if I had suffered some sort of psychological trauma of my own.

Meanwhile, every day, I watched TV and read print news reports of what was happening in Iraq and Afghanistan. I kept thinking about what my interview subjects had suffered and how the same terrible things were happening again, but this time to young American men and women of the armed forces as well as to civilians in Iraq and Afghanistan. I kept asking myself: For what? We had been told there were weapons of mass destruction in Iraq, which, of course, there were not. It became clear to people of all political complexions that this was a war for oil.

At the same time, that very oil, when burned to fuel our cars and make our electricity, caused massive destruction to our habitat's ability to support us. I saw college students walking around in shorts and T-shirts in January in New York City when it should have been twenty degrees Fahrenheit—climate change. I read more and more about the cliff the human species marched toward as we burned the exact fossil fuels we fight wars to get access to.

Nor could I get out of my mind that despite all our so-called progress as a species—the reason for which, supposedly, we burned all that fuel—one billion people on the planet didn't even have access to clean drinking water. I kept hearing about that "progress"—seemingly measured by how many television shows we could store on our cell phones—and wondering, *Wouldn't a better measure of progress be the kindness and compassion with which we treated each other? Or how well we took care of people and societies who are struggling?*

At the same time, I looked around me and saw that even the people who were supposed to benefit from all this fossil-fueled "progress"—the Americans and northern Europeans—weren't truly happy. Even my well-off New York City friends worked twelve- or fourteen-hour days and talked about which psychiatrist could prescribe the best cocktail of first, antidepressants, and second, stimulants to restore the sex drive the antidepressants took away.

I was deeply troubled about the world, and I couldn't shake the feeling that the complacent, consumerist life I led contributed to the problems. Nor was I getting rich or socking away that retirement I felt I had been promised. I was doing okay—privileged, even—but I still felt as so many of my "successful" friends did: constantly stressed, anxious, empty, and, frankly, unhappy.

What was the point of continuing to try to fit into the societal mold if it neither helped the world nor made me happy? I kept wondering, *What if I stopped following societal directions and pressure and lived according to my passions, my values, and my world concerns? What if, instead of trying to get something on the outside to make me feel better on the inside, I actually listened to the clues on the inside about how to make my external world better?*

That was the process that led me to launch No Impact Man, the project where I would make a number of transformational adjustments to my lifestyle (some deliberately extreme) in order to live more in accord with my values.

Back then, when I first launched the No Impact Man year of living as environmentally as possible in New York City, I made the critical error of thinking that living a life that was good for the world meant sacrificing a life that was good for me. I didn't quite think, as Colbert joked, that by "not poisoning the earth" I would be "poisoning our capitalist society," but I did think I would be poisoning my own comfortable way of life.

After all, when I started getting around only by bike to prevent fossil fuel pollution, I had to "give up" cars and other forms of motorized transport. When I started eating only locally farmed food to avoid shipping-related climate emissions and habitat-destroying agriculture chemicals, I "gave up" all those tasty prepackaged, processed goodies.

Pretty quickly, though, people started saying that the whites of my eyes looked brighter, my skin seemed healthier, my hair was thicker. I had more energy, I felt better, and I lost the paunch the doctor had been warning me about for years. Apparently, that's what happens when exercise becomes part of your daily routine—without even having to go to the gym—and you start eating food that is actually good for you.

Meanwhile, buying less stuff—another adaptation—meant needing to earn less money and having more time to hang out with friends. Eating eggs produced by farmers I knew to be humane, I didn't munch my way through breakfast worrying about the chickens. Making less trash, I didn't have those micro-twinges of guilt every time I threw something away. Visiting my congressional representative to encourage action on climate change meant I felt less like a victim of circumstance.

I had begun the No Impact Man project thinking that to live in line with my values meant to "sacrifice." But by a process of continual experimentation and adjustment—by actually being awake to my life, by being truly alive to my options—the list of better-for-the-world-and-better-for-me adaptations grew and grew. I finally realized that to *not* live in line with my values is the real sacrifice.

I'm not saying everyone should live the way I did during that year (I don't, though I maintain a fair number of the practices I developed). But what has become clear to me, from both my own experience and those of the thousands of seekers I have since been in conversation with, is this: even small efforts to live in line with our values, passions, and concerns offer a path toward a life that can be not only better for the world but also better for us. As so much ancient and modern wisdom teaches us, there is an unbreakable connection between being True to your Self and True to the world.

Learning to trust and apply the principle of being guided by values, passions, and concerns in all manner of life decisions is the central theme of this book. As you read, you will encounter and hopefully be inspired by the many examples of how other people have made authentic, passionate, service-minded life choices that made them happier and helped the world in every realm, from finance to friendship. You'll also find guidance, wisdom, and practical exercises to help you find your way to choices that are True for you.

As I've said, I'm not just talking about environmental lifestyle adaptations. If so many of the standard life approaches I'd challenged during No Impact Man weren't right, what about all the others? Say, in the realms of relationships, housing, home ownership, sex, childbearing, shopping, banking, retirement, friendship, education, spending, investing, citizenship, and the many other ways we relate to the world?

How do you live your calling? What career should you pursue? And how important is "career" these days? How much money do you actually need to live a good life? How do you make friends, and what really counts as community and family? Should you get married? Should you have kids, and if so, how many? How can you influence the policies of the society you live in? Should you really put yourself hundreds of thousands of dollars in debt for college? What does retirement planning mean in today's world?

So many of us have struggled to squeeze ourselves into our parents' and grandparents' generations' prepackaged answers to these questions. So many of us try so hard to follow standard approaches to life that were laid down back in the days before the Internet or even cell phones; in a time of Cold War, rapid industrialization, a predictable economy, and a planet with seemingly unlimited resources; and amid an array of repressive gender, race, sexuality, and other social norms.

But now, we all have a chance to join the quiet movement of freethinkers who are waking up to alternatives. When it comes to economic systems, for example, revolutionary thinkers have broken away from the false choice between capitalism and socialism by pioneering sharing and bartering networks. When it comes to the child-rearing dilemma, groups of people are starting to raise kids together through extended families, made families, mentoring, and other means. Still others are making strides in everything from farming to finance.

What are the benchmarks by which we can decide what is making us happier? What are the benchmarks by which we can decide what is helping our communities and the world? Is there a set of principles that can be applied in all situations? Are there shortcuts to navigating these decisions?

This book is among the first to mine and make public the tactics of the growing subculture of people who are finding new answers to these questions. Here is what they all say about the great unraveling: the amazing part is putting it all back together exactly the way you want to with the knowledge that all of us can take steps toward a happier life that also help build a much happier world.

The Truth About "Success" and Why This Is Not So Much a Self-Help Book as an Each-Other Help Book

Think back for a moment to my story about being on *The Colbert Report*. If this was your standard self-help book, and if I were sticking to the form, I would have told that story very much the way I did: with a vignette that establishes my wonderful so-called success. Because in the introduction of a typical self-help book, one of the things you have to do is convince your reader that you have mastered the standard life approaches in order to achieve the standard definitions of success—like being on TV.

The thing is, for so many of us, the time of "self-help" has passed. Think for a minute of the apocryphal story of the man who keeps trying to get the best deck chair spot while cruising on the *Titanic*. Getting the best chair spot on a sinking ship is, metaphorically speaking, like chasing after conventional success in a world that seems to be falling apart.

The reason the *Titanic* deck chair story is both so funny and so instructive is that when the ship is going down, having the best deck chair—winning conventional success—isn't going to do you much good. When the ship is sinking isn't the time to be thinking of deck chairs—or houses or careers or romance. The question is, how do you work with everyone else to keep the boat afloat? When the ship is sinking isn't the time for self-help. It is the time for each-other help.

When you think that way—of each-other help instead of self-help—your definition of success naturally changes. All of a sudden, making a big splash and getting semifamous on TV doesn't seem so successful anymore. In fact, it almost seems a little vapid and shallow. When you think in terms of each-other help, you have much deeper and wider and more satisfying ideas of what success means.

The truth is, I told you the story of my being on *The Colbert Report* as an attention-grabber. Not because I think TV appearances define me as successful. What credibility I have comes not from TV or other external circumstances, but from having finally learned that there are so many much easier and more important paths to real and actual success.

Here are some of the types of statements authors of self-help books sometimes use to establish their "success" and their credentials. Just to be clear, I am almost uniformly *unqualified* in those standard ways, and this each-other help book will *not* help you to achieve these goals, though it may help you achieve much better ones.

So-Called Self-Help Success: I have a $4 million house, and you can, too!

For the record, I most certainly don't have a $4 million house. In fact, I don't own a house at all. I rent an apartment. On a charming street in an edgy neighborhood that I love on the border of Clinton Hill and Bedford-Stuyvesant in Brooklyn. It's ethnically diverse and filled with students, queers, artists, writers, and, well, people.

My most recent downstairs neighbors, a straight couple, work in book publishing. The woman is British and the man is American. An old Jamaican man owns the house next door. He gets drunk sometimes and talks about the daughter he so sadly lost many years ago. There is a lovely woman named Anna who rents the first-floor apartment from him. She is a photographer. Two doors down are a hip-hop DJ and his family. Across the street lives a little girl named Emma who is kind enough to play with my daughter, Bella, even though Bella is three years younger.

My point here is, I don't have a $4 million house, and I don't think it is worth what it would take to try to get one.

Real each-other help success: *By knowing and liking my neighbors, I participate in a community that makes life better for both me and them.*

So-Called Self-Help Success: I've outsourced everything and reduced the time I spend working to almost zero, and you can, too!

Um, well, no. I haven't done that, either. I mean, here I am writing this book, which takes a lot of time, right? Earlier this week, I had to take time out to write an op-ed for the New York *Daily News* about how beneficial I think New York City's bike-share program is. Before that, I spent two weeks in Hungary, Poland, and the Czech Republic giving talks and running workshops on the quest for a better life.

On the other hand, after lunch today, I'll ride my bike over to Park Slope to have coffee with a friend. Yesterday, I took the morning off to attend my daughter's last-day-of-school assembly (she came with me on that trip to Central Europe, by the way).

What am I saying? I do have lots of work and other obligations that claim my time. I am not completely free. But the things I am obliged to do are things I *want* to do.

Real each-other help success: *I spend my time doing what matters to me, and I live my life in line with my values, which means I'm part of building a better world.*

So-Called Self-Help Success: I socialize with CEOs of major corporations and many Hollywood stars, and you can, too!

I won't pretend I don't know and correspond with and, yes, sometimes hang out with people who are well-known in my field. A couple of famous environmentalists. Sometimes a documentary filmmaker who works on social issues. Some writers. Some artists.

But when, in my past, I made the mistake of hanging out with such people because of their status instead of because of true bonds of affection, it was never very comfortable. Some sort of fame or standard-type success comparison would be triggered and everyone seemed to feel a little insecure.

Want to know some hanging out that is more important to me?

The night before last, I had dinner with a new friend named Kathleen whom I bonded with after her father suddenly died of a heart attack. Dinner started at seven and ended at ten thirty. We laughed and told secrets. I showed her my gold tooth. Tonight I'm going to hang out with another friend who has two kids and, like me, lives separately from his co-parent. We'll shoot the shit. Maybe order pizza. He certainly doesn't own a private jet.

So, no, I don't really hang with the rich and famous—at least not for that reason—but I have found something more important. I get to spend time with people whom I love and who enrich my life with friendship and trust

Real each-other help success: *I love many people, they love me, and we make time for each other to the benefit of all of our lives.*

As long as we have started this sort of inventory of what I think makes my life successful and what qualifies me to write an each-other help book, we might as well complete the list. Over the course of many years of questing, I've discovered that the life I'd rather strive for and often achieve is one that:

Lets me spend my time doing what really matters to me. (I don't have to work less because I care deeply about my work.)

Is successful according to my own definition. (It reflects who I really am, what I'm really good at, and what I really want—and a $4 million house ain't it.)

Keeps me safe. (Without being rich, I am taken care of, often in ways that have nothing to do with money.)

Contains a lot of love. (I have ample time for friends and those I think of as my family.)

Allows me to feel that I am contributing. (I never worry that I'm ignoring or am callous to the world's problems.)

Is full of fun and adventure. (New things happen all the time).

Integrates taking care of my body and my soul. (So I don't have to fit this in between other things I'm supposed to do.)

Includes taking time and space to explore my place in the universe. (Call that spirituality if you like.)

Allows me to do these things not by building large cash reserves but by building a life where what is important to me remains central so I don't need money to obtain it.

Helps me relate to the world—through my work, my relationships, my life choices, my living arrangements, my purchases, my civic participation—in ways that contribute to solving our global crises.

That's my list of what defines real success. It's not everybody's list, but it's mine.

Now here are the two big questions this book is going to help answer for you:

What is your definition of real success?

How can you make it a reality?

How This Book Works and Why You Don't Have to Change Your Whole Life to Make It Better for You

Maybe you want to be a vegetarian in a family of meat-eaters. Or maybe you want to do something more radical, such as move to Africa to volunteer in developing communities. Whether your goals are large or small, you can use this book to help find the way. It can be your launching pad if you are just starting, or your companion guide if you've already begun to build the life you want.

And remember, whatever our life circumstances, whatever our constraints, we all have a role to play in the world.

The peaceful revolution is composed both of people doing no more than growing tomatoes on their windowsills or tithing part of their salaries and of people doing no less than protesting in the street or upending their entire careers. The most important thing to remember is that you don't have to be a radical—though, of course, you can be—to apply the principles in this book.

The world doesn't just need local farmers; it needs the farmers' customers. It doesn't just need urban environmentalists; it needs the people who take food scraps to their compost piles. It doesn't just need the people starting bike-share programs; it needs the people riding the bikes. It doesn't just need race activists marching on the streets; it needs people retweeting the videos they send out.

One thing I realized, and you should, too, is that no one else's path will fit your life. As much as I've yearned for straightforward directions, there is no one person who has it right in every way so that I can just follow her lead. There are as many approaches to the new life as there are people in the world. What makes an extrovert happy makes an introvert miserable. How a city dweller can help the world is different from how a suburbanite can.

Following directions just means we end up where other people think we should go. Besides, how many of the world's problems have been caused by people abandoning their wisdom and assuming other people know better?

This book will help you to reexamine your life and truly take charge of it, not turn it over to another expert. It's about being awake to our moment-to-moment decisions, not about developing a new list of standard life approaches. It's not about how to become more like anyone else. It's about how to become more like you, and how to trust that being truly like you will be better for your life and better for the world.

For all these reasons, this book will not give you answers but instead will lead you through questions that will help you find answers of your own. Most important, it will help you find the compass points of your life—your passions, your concerns, and your values. Then it will help you decide for yourself how you want to manifest them in the range of relationships you have with the world.

To support that process, in addition to the questers we will meet, I will present you with both some surprising new science and some comforting but forgotten old principles that can help keep us on our paths, like guardrails by the side of the highway. Emerson's principle of self-reliance, Gandhi's thoughts on doing things for yourself, Tolstoy's thoughts on "resist not evil"—these are among the many principles we will examine as potential routes to the Good Life.

How you apply those facts and principles is an individual matter. So in addition to the real-life examples, the science, and the principles, I have included lots of exercises to help you understand yourself, your situation, your concerns in the world, what you know you need in life, and how you want to help the world.

In short, this book will:

→ Unpack the standard life approaches to ask if they are good for the world or good for you.

→ Look at scientific research and wise principles that might point to alternative choices that can make both you and the world happier.

→ Examine personal stories of people who have made new choices to see how new thinking can be applied in real life.

→ Provide exercises that help you see who you are in relation to these ideas and whether and how you want to apply them to your life.

→ Apply this approach to all the many relationships to the world that make up your life, from eating to parenting to investing to citizenship to forming friendships.

→ Help you make this process your own.

Whether you want to make any changes or not, this book will help you know who you are and what you believe. It's not a book to simply read. It's a work in progress. *Your* work in progress. It's meant to be thrown in your bag. Left on your bedside table. Set on top of the papers on your desk. Always within easy reach.

How should you live? How to be alive? There is so much to think about. In the end, this book should be more about what you've underlined, what you've written in the margins, and what you've starred than what has been printed in its pages. You'll want to work through it again and again in different ways.

Who are you? What, in general, makes you feel happy, satisfied, and safe? What makes you feel as though you matter and make a difference? The answer changes according to your situation. What is the situation? What is your relationship to the situation? Life is a series of small and large decisions. How should you make them?

The world is changing so fast, and as we have said, this is great news. You are going to get to rethink all the old decisions and find your own True Path. You get to make up your own rules. Decide what a successful life means to you. And then beat a path toward that life in a way that helps the world.

Frequently Asked Questions to Help You Decide if You Are Ready for the Ideas in This Book

Are the lessons of this movement of trailblazers right for me?

If you are looking for a life where you feel you are helping with instead of ignoring the problems of the world that get you down, then yes, these lessons will help you. Everything in this book points to life choices at the intersection of what makes you happy and safe and what helps the world. It is about being true to your own values, not someone else's.

Are you going to tell me there's something wrong with my chosen career and that I have to change it?

Actually, I'm not going to tell you that you *should* do anything. Change your career if you want to, or don't. But stop following other people's directions. It's time to put an end to a way of life based on what we *should* do. Look where it's getting us. Besides, adjusting your life does not necessarily mean changing your career. Career can't be everything anymore. There is so much more to all of our lives than just our jobs.

Does caring about the world and working to have a happy relationship to it mean I'm going to feel guilty all the time and be endlessly wringing my hands?

This book isn't about complaining. It's about fixing. Fixing your life. Fixing the planet. Work toward the good; don't focus on the bad. True fact: studies show that people who are actively working to make their lives and the world better have way more positive attitudes and are, yes, happier.

Do I have to be a hippie or a liberal or a conservative or a back-to-the-lander or have a particular life philosophy to make this kind of change?

No. All you have to believe—or at least *want* to believe—is that you matter to the world and the world matters to you.

Are you going to tell me I have to make sacrifices?

Fact of life: unhappy people don't make other people happy or help the world. They tend to be more self-centered. We're going to assume that the best life is one where making yourself happy makes other people happy and making other people happy makes you happy. So no, it is not about sacrifices. You may wind up making *changes* in your life, but if they bring you happiness, you're not likely to see them as sacrifices.

What if I'm already helping the world and I'm sick of being poor as a result?

See previous answer. If you want to *keep* helping the world, you are going to have to focus on your financial situation before you run out of steam.

Are you saying I can't have or want to have money in order to be happy?

No. But you might want to ask how much you are paying in life energy—time, relationship loss, stress, fatigue—to get your money. Because in some areas of life, there are much more effective and less costly ways to get the things you want than to buy them. So this book will help you figure out how much money is enough, for you.

I'm confused. Is personal happiness the end goal or not?

As I've said, this is more an each-other help book than a self-help book. This is partly because it will be a tremendous challenge to maximize your personal happiness if you have to keep living with the fact that the world is in a tailspin. After all, doesn't "self-help" feel a bit like trying to run up an escalator that's going down?

The world crises have blessed us with a deeper understanding of the fact that we are all connected. Our own happiness depends on the world's happiness and vice versa. The question is, how do we work with that fact in the many decisions we make in our own lives? The promise is that, though it might take work, finding the answer will bring you a much safer, happier, and more fulfilling life.

How I Failed at All the Standard Life Approaches and (Kind Of) Figured Out the Purpose of (My) Life as a Result

GOING FOR THE MONEY

Born in New York City, I move at age eight with my mother and sister to a little coastal town called Westport, Massachusetts. Self-conscious about being one of the few kids there living in a single-parent home, and the hand-me-down clothes I wore as a result, I set out when summer comes to make my fortune going door to door doing odd jobs—mostly garden work. My main customer—because people don't hire eight-year-olds—is my maternal grandfather, who points out bits of grass that I have missed when I mow the lawn around his house. I get "lessons" about responsibility and hard work. The takeaway: I hate gardening.

In hopes of avoiding it, I turn to comic books and fall for one of those "make a fortune in your spare time" ads. I receive a crate of greeting cards to sell, learn the preprinted sales pitch, bang door-knockers and ring doorbells for an afternoon, get bored, and leave my mother to ship back the largely unsold inventory. Over the coming months, I try selling my catch off the town dock (turns out there is no market for eels), shining shoes (hard sell in a town full of farmers and fishermen), washing windows (tough with no car when houses are a quarter mile apart), and selling eggs (neighborhood dogs eat my chickens).

I'm a bust at both Standard Life Approach #1: Work Hard and Knuckle Down—and Standard Life Approach #2: Get Rich Quick. Am I already a flop?

CRYSTALS AND ALL THAT OTHER NEW AGE STUFF

Conveniently, I decide money is not the goal of life since there is no taking it with you. A favorite uncle's suicide makes this terribly poignant when I am twelve. I grow my hair long and suddenly get into hippie-style spirituality. Richard Bach's *Illusions: The Adventures of a Reluctant Messiah* has got me lying on my back trying to make clouds magically disappear. I read Carlos Castaneda's *The Teachings of Don Juan* and want to go on a shaman's vision quest.

I'm on to Standard Life Approach #3: Look for Spiritual Shortcuts to Get What You Want (what some people call spiritual materialism). Lacking access to Don Juan's hallucinogenic peyote cactus, I experiment with the consciousness-altering (not to mention addictive) properties of Bacardi rum.

THE IVY LEAGUE ROUTE

I spout the philosophy behind my so-called alternative path to advance my conventional path—an application essay entitled "Why

I Don't Want a Rolls-Royce" wins me a place at a fancy prep school. The idea is to get good grades, get into an Ivy League college, win a place in the right social circles, and set yourself up for life. This is Standard Life Approach #4: Go to the Right Schools.

The price for Approach #4 is that student debt and the credit card debt you accumulate while paying your student debt keep you tied to your "career path," whether you like it or not, until you graduate to old-age heart attacks, cruise ship vacations, or both. I'm spared this fate when the consciousness-altering properties of Bacardi become too fascinating. I get suspended and then kicked out of prep school. Also, I have a couple of minor but alarming brushes with the police. Sorry, Mom and Dad.

KNUCKLE DOWN, WORK HARD REDUX

Trouble scares me onto the straight and narrow—with help and support. I come away from the Bacardi predilection strongly aware that it wouldn't have taken too much bad luck for my life to have totally gone down the drain. I think of others not lucky enough to get help. I think of the accidents of birth and skin color—privilege—that meant my brushes with the police didn't destroy my future. Does my good fortune in getting the help and support I need belong to me alone? No. Even back then I realize—and am taught by good adults—that it is something I'm supposed to pass on.

Meanwhile, I back away from all pretenses of alternativism, cut my hair short, and wear button-down oxford shirts. Along with attending Westport's public high school, I'm back to odd jobs. I land a gig scraping and painting the front of a clapboard house, even though I hate painting nearly as much as yard work. But going back to my grandfather's lessons, hard work and responsibility are the key, right?

Somehow, I become a distributor for a type of cardboard toy airplane that does a loop-the-loop and lands back in your hand—the Dip-er-do—and travel to occasional fairs to sell them. The rest of that summer, I mow lawns in the morning—which, strangely, I enjoy this time around—and visit an old lady who pretends to need yard work done but really just pays me to chat. In the afternoon, I go to the beach. I earn enough to buy my first car.

I procrastinate and procrastinate painting the house. Finally, I feel guilty and get up really early one morning and wake the neighborhood applying the belt sander. The lady who owns the house comes out in her bathrobe, tells me I'm nuts, and fires me.

I am scared that I'm no good. I don't seem to be able to knuckle down. I believe that I'm a loser, while entirely missing the point that, contrary to my grandfather's lessons about working hard at what we *don't* want to do in order to thrive, I'm thriving at what I *do* want to do—the airplane sales, the lawn mowing, the conversations with the lady—and getting fired from what I *don't* want to do. This is an important lesson I'm not yet old enough to understand. Also, where has my intrigue with the get-rich-quick approach gone? Turns out it disappears in people when they have a good life that they enjoy.

FITTING MYSELF LIKE A SQUARE PEG IN A ROUND HOLE

Finishing up high school, I am so excited by all the things I could study at college that I can't make a choice. Crazy societal idea: pressuring eighteen-year-olds to decide what they want to do in the world before they even enter that world (like forcing someone to buy shoes without letting him try them on).

I'm thinking of astronomy. Stars! Where do they come from? Where do they go? What is the universe? How big is it? What made it? My grandfather listens to these existential questions patiently over lunch, then says, "You need to study something practical that will get you a job. Being yourself is great. But that's for your spare time. This is the time when you have to grow up."

Actually, this is the time when so many of us stunt our own growth. Because we decide that we are too scared to go our own way. We force ourselves to forget that life is magical. We learn to override ourselves. What makes sense is to create your life in line with who you are, what you love doing, and how you'd like to help the world. I don't yet trust myself to do that. I choose Standard Life Approach #5: Believe That Life Is Already Created and You Need to Fit Into It.

CONSOLATION PRIZES START FLOODING IN

Can't get more practical than electronic engineering, right? That's the major I choose. Having spent so many summers in the United Kingdom—Dad is British—I go to the University of Liverpool (also because it was then free—God love socialized education). Turns out I'm good at my field and graduate first in my specialty. Professors groom me for academia and offer me full pay to do a doctorate.

Part of "becoming a grown-up" is moving out of communal living in college dorms—probably the happiest part of my college experience (an important lesson for later)—and into my own apartment. I'm miserably lonely.

Some key information about myself I've overlooked because I'm still trying standard life approaches: I love beach and leisure time, am very social, believe my good fortune should be spread to the world, like variety and have an entrepreneurial bent, am a bit alternative by nature, and don't care that much about money.

Here is what fitting myself in gets me: good pay and a guaranteed life of office swivel chairs working alone in a lab. Not actually what I wanted. See what happens when you override yourself?

DEPRESSION

Life is feeling like a dead end. I end up in therapy. Visiting my father in Israel, where he is working, I watch the red glow of his cigarette on a dark late-night drive. *What is life for?* I ask him. I am wrestling with so much. Do I exist for myself or for others? Should I be a social worker or an entrepreneur, a monk or a merchant? What life will I feel best about having lived when I die? Can I make a sufficient contribution to the world to earn a happy place in it? What actually counts as a contribution?

My misery has thrown up all the existential questions. But people keep telling me to ignore the big questions (Standard Life Approach #6). You'll never answer them, they say. Is that what we are really supposed to do? Put life's biggest questions out of our minds? Can I be sure that life isn't about the opposite—investigating them?

Besides, without fundamentally understanding my nature—what am I?—how do I know what my life is for? How are any of us supposed to live? Should we really follow the Standard Approaches to Life even if they seem to make us deathly depressed?

FROM BAD TO WORSE

My Ph.D. supervisor bursts into the lab all excited. There is an application for our work that can help ensure that NATO submarines carrying intercontinental ballistic missiles can't be detected. Think of the Department of Defense funding!

Nuclear bombs?! That's the last straw. Is that really my life path? Is that really anyone's life path? Depression and destruction? I quit. I'm now a Ph.D. in a field I've decided to leave forever. Lovely!

I buy a book called *The 100 Best Companies to Work for in the UK,* skip those companies' recruiting departments, and write directly to the CEOs. "I don't want to work for your company. I want to work for you!" I say. The CEO of Iceland Frozen Foods, a seven-hundred-store supermarket chain, hires me as his personal protégé.

Executive suite at twenty-six. I'm a life-hacker, right? Sadly, no. I'm still just career-hacking. Real life-hacking doesn't come to me until later, when I realize the wrongness of Standard Life Approach #7: Go All Out for an Amazing Career and You'll Get an Amazing Life. Jealously, I say good-bye to a bunch of friends who go off to live in Africa, basically for the hell of it.

BREAKTHROUGH #1

I spend a day with the company security chief learning to spot shoplifters and installing secret cameras to spy on managers suspected of stealing. It leaves a bad taste in my mouth. Two months in, I realize

I've taken another wrong turn. I'm not using the skills I want to use. I'm not finding meaning. I'm learning that I can't fit myself into the world the way my grandfather told me I should, or at least I don't want to. I quit, and the CEO calls me a twat.

I put in a miserable call for help to a headhunter I know. He puts me on hold, comes back a few minutes later, and tells me I have an interview the next day. By the end of the week, I am, crazily, the second-in-command of a boutique public relations firm. I'm suddenly writing and speaking and communicating and doing things I naturally enjoy—almost as fun as selling Dip-er-do airplanes and mowing lawns (though I don't get to go to the beach in the afternoon).

One problem: I don't care one iota about the missions of our clients—a bank, an electric hand-dryer maker, etc. On a trip to Ireland for an equipment manufacturer, I end up at a slaughterhouse that the company has kitted out. Pig carcasses hang from the overhead assembly line and then tumble in a huge clothes-dryer-like contraption that burns off their hair. My boss and I are vegetarians. What the hell are we doing?

BREAKTHROUGH #2

Economic domino effect: About a year after I start working there, my boss's company goes belly-up when three companies that owe it money go bankrupt themselves. The one client I like, a charity that provides low-income housing, says it will stay with me if I go out on my own.

Meredith Beavan Publicity is born (Meredith is my middle name). It provides public relations services exclusively to not-for-profit clients—social housing providers, HIV prevention groups, a free therapy clinic, a couple of nonprofit hospitals. I have a sense of meaning, collaborate with amazing people, am self-directed, use many of the skills I care about, and learn a huge amount.

Not only that, but I have the leisure time I need and many good friends. For the first time, I am, as they say, doing well by doing good. Four great years pass.

MAYBE CAREER ISN'T EVERYTHING?

I keep having this daydream about riding around the United States on a motorcycle. I actually figure out a way that I can take a three-month sabbatical, but it scares me. I have this sense that if I do it, I will never come back. My life will unravel. Guess what? I'm right. But in a good way. After three months zooming around, I go back to England, close everything down, pack up my life, and move to the United States.

I have no plan. I am lost again. What do I want to do? I refuse to simply relocate my career based on my qualifications alone. Is that really the way to choose your work life? Do what you know how to do regardless of whether it makes you or the world happier? Another Standard Life Approach (#8) I am not interested in: Do Only What You Are Qualified to Do. I want to choose. I want to be proactive. I resolve to wait until an answer appears.

TRUST THE PROCESS?!

For two years, I split my time between waiting tables in Providence, Rhode Island, in the winter and driving taxis on Martha's Vineyard in the summer. I feel amazingly free and I am scared shitless. Meanwhile, I write countless short stories in my spare time. I submit them to magazines and paper my wall with rejection slips. Being a writer is a childhood dream.

I start thinking of moving to New York to really try the writing thing. Two questions plague me: First, *How do I know I can be a success?* Second, *What good would my being a writer be to the world?* I ask this question because being of service grew to be an important value for me during my years of PR.

I agonize. One day, I'm sitting and meditating and I literally hear a voice say, "It's okay to desire things for yourself if you point the energy of that desire in a direction that will help others." This is a hugely important principle, one that I would learn the Buddhists call "using your karma to help the world." I decide to trust that my being a writer will one day help people, at least in some small way.

FOLLOWING MY PASSION AT LAST

I finally have found a path I *want* to follow rather than one I should or simply could. I know what I want to do with my life and throw myself in. I quickly get work writing self-help columns for *Esquire* and other magazines like *Men's Health*. Before long, I break out of self-help and write articles for *Atlantic Monthly*, the *New York Times Magazine*, and others.

I get a contract to write my first book, *Fingerprints*, about a nineteenth-century murder case that was the first solved using fingerprints. Then I get a mid-six-figure advance for a second book, *Operation Jedburgh*, about the Allied soldiers who dropped into occupied France during World War II, an operation my grandfather participated in (yes, the one with the fetish for hard work).

Following your passion. Expressing yourself. Being creative. The secret to life, right? I'm doing all that. I have, I guess, "arrived," except I am nagged by a sense of meaninglessness. I keep hearing that voice that said I could pursue what I wanted as long as it helped people. *When am I going to help people?*, part of me keeps asking. *Isn't there a path that isn't just about money or meaning or self or other but about all?*

The price of a good life, people keeping telling me, is that I have to follow Standard Life Approach #9: Accept the Reality That You Have to Earn a Living and Let Go of Childhood Idealism. If you're good at finance, you have to accept that the world of banking is corrupt and screws the little guy. If you're good at politics, you have to accept that Big Money does the voting, not the people. If you're a writer like me, you have to accept that most readers are really looking for a way to escape. It's just like television, except with an upturned nose.

You can't buck the system. You are too small. You are too powerless. By the same logic, it does not matter if you drive gas-guzzling SUVs or buy clothes made by Bangladeshi child laborers. Because what difference do your tiny purchasing decisions make anyway? What difference does your whole life make? You might as well just do whatever makes you happy.

Except this approach doesn't make me happy. I have to willfully ignore the world's problems, and so many people I know do, too. One of my best friends loves going to Africa to practice simple dentistry on people who are losing their teeth. He loves that, but he gets to do that for only a week each year. He spends the other fifty-one weeks ignoring the world's real dental problems and trying to convince rich people to shell out for cosmetic dentistry to improve their "smiles."

I do what so many career-driven New Yorkers do: I try to do good at the margins. The "philanthropic" approach. Do whatever you need to do to earn the money and then try to ameliorate the harm by giving some away. Doing bad to do good. In my own version I tithe to charity and volunteer in my "free time" (whatever that means when so many of us are working fifty, sixty, or seventy hours a week).

You could say I've "matured." Sooner or later we all have to realize that the idealism of youth is just that—idealism. But is that true, or is it just a story we tell ourselves to justify choices we are not really happy with? Is rejecting the idealism of youth just a way of closing our eyes to the violence of adulthood?

THE WORLD GOES SOUTH

The Iraq War breaks out. More than four thousand young American soldiers ultimately die. I obsess over a web page called Iraq Body Count. It eventually documents more than one hundred thousand civilians killed.

On the one hand, there is what we do for oil—the war, the BP disaster in the Gulf, etc.—and on the other, there is what happens when we burn it—climate change (not to mention the many other crises we now face in the Earth's ability to support the way we live). Between those two things is our way of life, the thing we use the energy for. With 25 percent of Americans suffering from depression or major anxiety, I wonder, *Is it all even worth it?*

I can't ignore that voice anymore. I can't go on writing history books. Instead, I launch what I call the No Impact Man project.

At first I believe that the project is about the repudiation of desire. That what is making people so unhappy and destroying our habitat is the human desire for more. I am trying to be an ascetic and dragging my family along on a quest for less. It takes me some time to realize that the rejection of desire is yet another standard life approach, an automatic, simplistic reaction to the failure of the other ones.

Because of my project, though, I begin to meet other people who are investigating nonstandard life approaches and are far from ascetics. Instead, they are searching for happiness in nonstandard ways. From them I learn the value of reexamining not just career choices but all life choices.

I see that there are standard life approaches to everything: how we purchase, where we live, whom we have as friends, whether we have friends at all, how much we emphasize money and careers, whom we have sex with, what form romantic relationships should take, what we eat, when and how we retire, how we participate in the political process and on and on and on.

The truth is that many of these standard life approaches are failing, from both a planetary perspective and a simple human happiness perspective. Can we afford to buy everything ourselves? To own our own lawnmowers instead of sharing with our neighbors, to waste money and resources on a power drill that will get used for about nine minutes over the course of a lifetime?

The surprising result of No Impact Man is that I learn that we don't have to deprive ourselves for the sake of the world. It is not about not wanting things. Instead, we simply have to learn to want the things that actually make us happy. It's not about suppressing our desire but nurturing our wisdom.

Who am I? What are we? What do we understand about what really makes our lives worth living?

These are frightening questions, because to answer them we have to be awake—both to ourselves and to the world, to its dangers as well as to its opportunities. The nice thing about the standard life approaches is that they pretend to offer guaranteed success if you just suppress yourself and follow directions.

But with the joblessness and home repossessions of the world financial crisis, can any of us be sure the standard life approaches even work anymore? Not to mention that when we suppress ourselves we lose our inner guidance. To what extent have we sleepwalked into a crisis of ecology, economy, and existence? Good news: all we have to do is wake up to ourselves.

NOW THAT I'M LIKE YOU AGAIN

How do you become the person you want to be? How do you get to be happy in your work? How do you construct the rest of your life? How does one live a life doing more good than harm in the world? How do you live a happy life balancing your own needs with those of others? How do you follow your passion? How do you even figure out what your passion is so you can follow it?

How do you do all that while making yourself happy, taking care of your loved ones, and, somehow, addressing yourself to the world's problems? How do you do all that while recognizing that the world is rapidly changing? How do you figure out what is most important? How, in short, do you live?

These are the questions that have plagued me all my life. Here is the thing: There is no straightforward answer. There is no standard life approach that works for everyone. Many of the world's problems come from too many of us trying to follow directions, because many of the directions are not good for the world and not good for us. Which leads back to that great part: that there is an amazing opportunity for each of us to wake up to our own paths that both make us happier and make the world happier.

As the No Impact year ends, I am back to facing the same questions and dilemmas and tug-of-wars between desire and altruism as everyone else. The No Impact year, in some ways, was like living in a monastery. Now the year is over, and I'm back in the marketplace.

I do a lot of public speaking. I run an international nonprofit that teaches people to find answers to their own life questions. My family has changed form and I am no longer married to the mother of my daughter. I am older and facing retirement. I may not have the kind of older years that my grandfather had. On the other hand, I'm not sure he had the life satisfaction that I have. He didn't experiment the way I do. He lived in a different world.

What I've found in this world is that you can live an amazing life if you tear up the standard life approaches (notwithstanding limitations like physical ability, extreme poverty, denied access to resources and others, of course). Not that you have to tear up all of them. I certainly haven't. There are realities to face. This isn't the No Impact year. This is my life. Living morally in the monastery is easy. The real question of our times is: How do you live a moral, balanced, happy, fun life when you're in the marketplace?

Finding that path is the purpose of this book. It doesn't start by concentrating on the problems. It starts by looking at the goals, the aspirations, the things we could be. It doesn't start by worrying about how much less harm we should be doing but how much more good.

What do you have to give to the world? Are you ready to insist on giving your gifts? For me, my gift is my voice. But it isn't that for everyone. Each of us must find our own path. Slowly, slowly, I've come to find my place in the world. Moment by moment my life mission reveals itself. I hope this book helps you find yours, too.

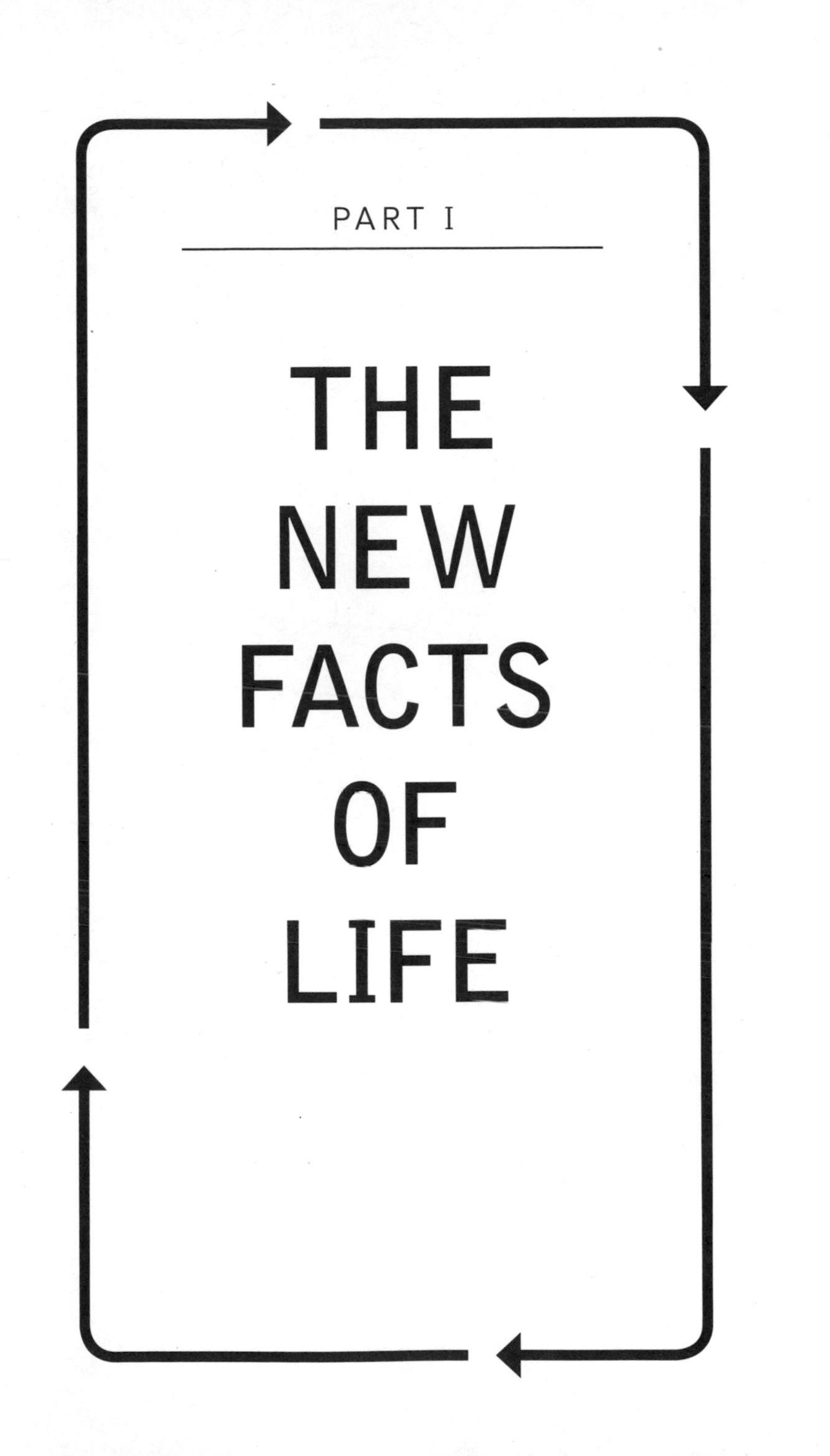

PART I

THE NEW FACTS OF LIFE

1

THE AWFUL STORIES WE TELL OURSELVES TO MAKE OURSELVES SMALL

life·quest·er
/līfkwestər/
noun
1. Someone who tries to understand her True Nature and uses that understanding to make a better life for herself and others. 2. A person who attempts to live his own values rather than society's and is excited by what he might discover. 3. Someone who has faith that letting go of limiting stories about herself and the world will allow something wonderful to open up. 4. A fun-loving, wicked cool person.

As we've discussed, the world is full of standard life approaches—cookie-cutter goals and paths that we feel compelled to pursue regardless of whether they truly make us happy. But what's the source of our compulsion? How does society reinforce these standard life approaches? The culture tells us stories about how the world works and how we fit into it, and we repeat those stories to ourselves over and over and even add to them and then call them the facts of life. We try to squeeze ourselves and our lives into those molds, even though, deep inside, we know we don't fit. To do that, we have to cut parts of ourselves off, turn down our volume.

Take Anita, for example, who told me her story on Facebook:

> *I was stuck in a job I didn't like. My sister-in-law had just died of cancer at age thirty-nine, which frightened me so much because my mother had also just gotten a diagnosis of cancer. I was scared and miserable and felt powerless.*
>
> *Every week, I went shopping at cheap chain stores like H&M, even though I didn't need any more clothes. I was just looking for distraction.*
>
> *I heard about the big garment factory collapse in Dhaka, Bangladesh, that killed over a thousand people. I knew the cheap clothes I was buying came from places like that and that the workers lived in miserable conditions.*
>
> *The thing is, I worked in the nonprofit world and I really cared about such social issues. But I pushed them to the back of my mind. I couldn't do anything about my sister-in-law. I couldn't do anything about my mother. And I certainly couldn't do anything about faraway garment workers. Those are the facts of life.*
>
> *But if those are the facts, why didn't I feel right about any of it? Why did I feel so guilty that I always hid my new clothes from my boyfriend as soon as I got home?*

Like Anita, we might push our concerns to the back of our minds and accept a life that we don't feel quite right about. How many of us feel we are stuck in a life that doesn't fit? One where we believe we can't fix anything, so we console ourselves by buying things we know we don't need? One where distracting ourselves from the rebellious feelings of despair that bubble up inside us becomes our goal?

That's the difference between what I call the "lifequester's" life approach and the standard life approaches. Lifequesters are people who have woken up to the possibility that they don't just have to try to fit themselves to the old so-called facts of life. Lifequesters are people who have begun or are ready to begin the search for the kind of life choices that are both true to themselves and true to the world.

To be a lifequester means to be willing to step away from the old and poisonous stories we tell ourselves that would otherwise force us to accept limited lives in a suffering world. Stories like the ones that underlie Anita's tale. Stories like:

- You don't have control over your life or the world, so you might as well find whatever you can to distract yourself from what troubles you about them.
- You are tiny in the scheme of things, so there is no point in changing your life to help fix the world's problems.
- Dealing with your own pain is hard enough, so don't overwhelm yourself by trying to assist with other people's pain, too.
- There is only so much happiness in the world, so you have to accept that your happiness comes at the cost of others'.
- And the biggest falsehood of all, the one upon which all the others are based: we are separate from each other.

The stories we tell ourselves about life are the posts and beams of the personal worlds we end up living in. Your stories are the illusions that make your personal Matrix appear real. If you tell yourself you can't climb off the treadmill, then you can't. If you tell yourself that nothing you do matters, then it won't.

Every time we tell ourselves a story about how we are less powerful than we truly are, we *become* less powerful than we truly are. Believing the story, we stop trying. We close down our capacities.

Sometimes we cling to stories that divorce us from our abilities because we are scared of what we would become if we accepted those abilities. Maybe we are scared of who we would alienate or whose denial we would challenge. Maybe we stay stuck because we know that so many of our relationships are based on believing the same stories.

Sometimes we cling to our stories because we fear there is nothing else—we fear emptiness. But our lives are comprised not of our stories but our experience. Even without the illusion of a story's security, basic facts remain true: The floor is still hard beneath your feet. The leaves on the trees still rustle in the wind. Sugar is still sweet. Salt is still salty. Reality is right here, whether you cling to your stories or not.

If you can detach from the stories you tell yourself about how you should or shouldn't relate to the world, then you will simply relate in the way that comes most naturally to you (what humanist psychologists call *organismically*). In the way that brings joy to you and transmits compassion to the world. In the way that helps you and helps others. To know this, all you have to do is stop believing you are less than you are and unbind yourself from your limiting stories.

The Science of Stories: Why the Limiting Stories We Tell Ourselves Sap Our Life Force, and the Good Things That Happen When We Let Them Go

First, our brains collect stories:

What I mean when I talk about "stories" is what cognitive psychologists call *mental models*. A mental model is a simplified representation of reality, stored in the brain, that helps us predict the consequences of our actions. Cognitive psychologists widely agree that people develop and use these representations as one of the bases of their decision making.

Our mental models—our stories—can be built from our own experiences, from other people's experiences, from our understanding of the world, from our opinions of how things should be, from what we believe other people expect from us, or from many other things.

Some mental models are not grounded in our experiences at all but are handed to us by our culture (through people, media, and even—God help us—advertising). Mental models handed to us by the culture form the basis of what I have been calling standard life approaches.

Then, as we go through life, we compare our experiences to our stories:

When we encounter a real situation, the brain flips through its story (or mental model) file until it finds a match. Mental models help us transfer learning from one situation to the next—through what psychologists call *analogical thinking*—so we can get faster and better at dealing with things.

So, say you're a kid and you see a barking dog. Your brain searches for a barking dog story file. Maybe your mom told you a story about how she got bitten by a barking dog when she was little, so you react accordingly.

Finally, our brains react to the stories (rather than reality):

A barking dog? Run!

In some ways, mental models are like bits of software that get loaded into our brain's hardware and executed when triggering events occur. Of course, those events can be much more complicated than a barking dog. Our stories affect how we react when we encounter Republicans or Democrats or Muslims or Christians or when world crises impinge on the happiness of our own lives.

If we aren't careful, we can end up living more in our stories than in reality:

Each person's brain collects thousands and thousands of mental models—stories. Taken together, the web of stories forms not just a single piece of software but a whole virtual game world in which we live, our own personal Matrix.

As beneficial as mental models can be in predicting outcomes and helping us choose our behavior, we can forget to check our stories against what we see with our eyes and hear with our ears. We don't ask ourselves if our stories match the Truth. We can cling to the map and forget to watch the landscape.

Through the process psychologists call *denial,* we accept into experience only those aspects of reality that accord with our stories. We literally can't see the parts of the world that don't fit our stories. We don't see that the barking dog is also wagging its tail—a puppy who just wants to play. We can see only the barking dog that threatens to hurt us.

Then our real world begins to conform to our story world:

We don't choose our stories based on whether they are true or false. We choose them for the effect they have on us. Perhaps they make us feel safe (I know I'll be fine with enough money) or happy (everything turns out all right in the end) or justified (people aren't nice, so why should I be?).

Sometimes, we prefer the effect of the story to the reality of life. We like our little answers more than the big questions. But then, we are no longer free to act from our True Selves because we follow the dictates of the story. The story uses us to fulfill itself instead of our using the story to fulfill ourselves.

Because of our denial, we can't see and don't pet the barking dog who also wags its tail and wants to play, so the playful dogs ignore us and move on to someone else. Playful barking dogs stop appearing, not only in our story world but in our real world. We are left with only the barking dogs who mean to hurt us.

That is how our stories, when we are too attached to them, create the world we live in.

The big question: Is the world your stories are creating the one you actually want to live in?

For Anita, for example, her story of powerlessness over the plight of the garment workers may have offered comfort. We tell ourselves stories of personal powerlessness partly because if we acknowledge that we are powerful, then we must also acknowledge that each of us is in some way responsible for the world.

That brings difficult questions: What can I do about climate change? What can I do about world hunger? What can I do about racial inequality? What can I do about big money in politics? What can I do about my own happiness? These questions are scary. And because you can answer them only for yourself, they can make you lonely.

But not as lonely as you will be if you abandon your True Self and divorce yourself from your capacities and abilities. Without accepting your power, you are in the world without your Whole Self—the loneliest state of all.

When we hold on to our limiting stories, we slowly turn off aspects of ourselves, until we are like robots, chasing after this or avoiding that to try to numb ourselves to painful situations our stories tell us we cannot change. We get so lost in a net of self-imposed obstacles that we no longer know how to be alive. The only response we are left with is to go shopping for clothes we don't even need or watch TV or keep showing up at the job we hate.

You cannot selectively close down your ability to respond. You can't be half-awake and half-asleep. Turning off to the world means turning off to our Selves. When we deny our ability to help the world, we deny our ability to help ourselves.

Now listen to what happened to Anita when she let go of her stories and accepted her ability to respond:

> *One day, I realized I wasn't hiding my new clothes from my boyfriend because of the face he made when I brought more unnecessary clothes home. It was because I didn't feel right about it myself, and my boyfriend's reaction just reminded me of it.*
>
> *In my heart, I knew the shopping wasn't good for me and it wasn't good for the world. I was betraying my own sense of values but I felt I deserved to do the shopping because I deserved the distraction given what had happened to my sister-in-law and mother. What I didn't realize, though, was that the only person I was rebelling against was myself—which meant I was actually trying to make myself feel better by making myself feel bad!!!*

I'd been pushing parts of myself down for so long that I felt dead inside. I felt suppressed. Like I had no sense of creativity. I felt so lost in a net of obstacles that I no longer knew how to be alive. That's exactly what I didn't want from my life.

I didn't know quite what to do but I knew that shopping was a big part of the problem. I had strong values and I was ignoring them. So I thought, what if I did the opposite? What if I embraced the truth of my values? What if I decided to not participate in what I felt was bad for me and bad for the world? What would happen then?

I decided not to shop for a year. No clothes, no shoes, no accessories, no bags. Since I wasn't participating in the global garment trade, I was also able to inform myself without feeling bad about what I found out.

I read about the appalling labor conditions and saw videos of women who had lost a limb because of accidents during work. I learned about the environmental catastrophe caused by conventional cotton farming (one-third of all pesticides worldwide are used on cotton plants). And of course I was appalled by what was happening to the women in the garment factories of Bangladesh.

I couldn't go back to my old shopping habits after this year. On the other hand, I still had to find a way to get dressed in the morning.

So I decided I would look for alternatives. How could I dress myself well without participating in a system I didn't believe in? I began investigating free trade and secondhand and ecological brands. I started blogging about all this and I really enjoyed it.

It turns out that people were really struggling with the same things that I struggled with and they were looking for guidance, too. Before long, a publishing company approached me and now I'm writing a book.

Since no one story is Ultimately True, the only question about our stories is whether they assist our happiness and ability to function in the world. If you make a particular world, you live in that world. Do you like the personal world your stories make for you? Is there a better world you would choose to make for yourself and others?

Those questions are for you to answer. But here is one key:

For a fulfilling life, any story you choose must include an acceptance of your power over and responsibility for your life and the world around you. That may not be true for everyone, but it is certainly true for me and probably true for you, too—since you have found your way to this book and read this far.

Accepting power and responsibility simply means accepting your *response ability*—your ability to respond to the problems in your life and in the world. It doesn't mean you are to blame. It doesn't mean you have to fix everything by yourself. It doesn't mean you have to figure out how to control the results. It simply means you have to *stop confusing not knowing how to respond with not being able to respond.*

Not knowing quite how to respond is okay. Telling yourself you can't respond is not. This is why the ancient Taoist text the Tao Te Ching asks, "Can you remain unmoving until the right action arises by itself?" Don't worry. Just choose for yourself this story: "The answers will come." Finding them for yourself is what this book is about.

When we stop binding ourselves up and limiting ourselves and hiding our light under our old stories, our action emerges naturally from our True Values and our True Selves. We have a better chance of helping ourselves and the world.

When we choose to live according to our values, people pay attention and the ripples spread. When we take responsibility for the world's problems—accept our ability to respond—others are inspired to do so, too. We feel as though we matter. We get happier lives.

Paint a Bull's-Eye on Your Limiting Stories

1. Sometimes the limiting stories are so old and so deep that we live in them the way a fish lives in water and doesn't even realize it's there. Writing down the answers to these questions will help you identify some of the stories you are living in. You don't need to challenge them. Just write them down.

a. Write the names of three people you consider heroic. What did or do they have in their character that you believe you don't?

b. What obstacles in the world and in yourself stand in the way of your changing your life in ways you want to?

c. What stops you from making a difference in the issues you care about in the world?

2. Now, recast the answers above by saying explicitly that they are stories you tell yourself. So:

a. Write "I stop myself from being more like [*hero's name*] because I tell myself a story about how I don't have ________________." Repeat this for each characteristic you said you don't have for each hero.

b. Write "I stop myself from changing my life the way I want to by telling myself a story about ________________." Repeat this for each sentence you wrote down above.

c. Write "I stop myself from making a difference to the things I care about in the world by telling myself a story about ________________." Repeat.

In the future, when your inner voice tells you a limiting story, repeat it to yourself a second time, adding "I have a story about . . ." at the beginning of the thought. You don't have to decide whether or not the story is true. Just realizing that it is a story is a big step forward and paints a bull's-eye on your stories for our later work.

Three Poisonous Cultural Stories Designed to Keep You Asleep and How to Break Away from Them and Any Other Harmful Story

So:

Underneath so much human suffering is a false story about the world. Too often, we don't suffer because of what is, we suffer because of our stories about what is. If we dig out the story and show ourselves how it is false, then the story loses power over us and we are free to become the people we would be without the story. (That, by the way, summarizes Buddhism's Four Noble Truths and the two Judeo-Christian commandments regarding false idols.)

Of course, there are all sorts of stories that cause all sorts of suffering, and the technique we are going to discuss here can be used to disempower all of them. But to demonstrate the method, I am going to examine three groups of similarly themed cultural stories that would-be lifequesters often grapple with—stories that make the search for less conventional life choices that are better for both the individual and the world seem pointless.

As I've traveled the United States and the world, and talked to thousands and thousands of people, these three groups of paralyzing stories are the ones I have most often helped people defang for themselves. Nobody wants these types of stories to be true. They don't make anyone happy. They block our light and make us feel powerless, but because so many people believe them, it is hard to trust ourselves enough to acknowledge that they're false.

Paralyzing Story #1: The Unbeatable Enemy Within

People are too selfish/dumb/unevolved/flawed and there is nothing anyone can do to change human nature, so trying to help the world is pointless.

The basic theme of this set of stories is that fundamental flaws in human nature ultimately doom us. We are either too selfish or too unconscious or too dumb or too what have you.

As individuals, therefore, each of us might as well just follow the rules and directions. If we are arrogant enough to think we can find a different path, we will personally fall victim to those same flaws that undermine humanity. Also, trying to help humanity as a whole is fruitless because our very nature damns us and can't be changed. Lifequesting, according to this set of stories, is pointless.

Is it really true?

This is the first of a series of questions that constitute "The Work," a system of identifying and unpacking our false stories developed by the self-help guru Byron Katie. In her system, you ask a series of questions to help loosen the grip a story has on you so that something more authentic can emerge.

Applying this first question in the case of Paralyzing Story #1, we ask, is it really true that people are too selfish, for example, and that you cannot change human nature, and thus trying to help the world is pointless?

The question, by the way, is not "Did someone teach you that it is true?" or "Does our economic system assume it's true?" or "Do you sometimes experience it as true?" The question is "Is it Absolutely True?" Continuing with the people-are-selfish version of the flawed-nature story, is it Absolutely True that all people (including you) in all situations always act selfishly? Is it Absolutely True that selfishness is at the root of human nature—and of your nature—and people can never be persuaded to behave otherwise?

Have you ever seen people act less selfishly than they did before? Have you ever seen anyone give somebody something without the expectation of reward? If so, can you really say that this selfishness story—or whatever other story you are applying Katie's questions to—is Absolutely True?

Not long ago, I saw a homeless man helping an elderly lady cross the street. If we were always selfish, shouldn't the homeless man be mugging the old lady? Shouldn't the old lady be shooing away the homeless man?

Therefore, can we know for sure, can we positively prove that it is Absolutely True that people are selfish? No. Even if you can't find a personal experience that casts doubt on the story you are working on, is it possible that your experience of the world is limited and therefore not Absolutely True?

Our stories are never Absolute Truth. The answer to Katie's question, *Is it really true?*, is always no.

Next ask, *How do I react—what happens—when I believe this story?*

First, at certain times in my own life, the "people are selfish" story has forced me to live in a world where I believed people behaved like animals fighting for scraps. If I wanted to get ahead, then I would have to fight for scraps, too. The story of selfishness makes me feel sad.

Second, believing people are selfish forced me to distrust the reality of what I saw when people behaved generously, which caused me a lot of anxiety. What I actually experience, for the most part, is kindness. I see that most of us behave as though we want the best for our neighbors and our coworkers. How could I possibly not be anxious about trying to reconcile this actual experience of people with my story of their selfishness?

Third, the selfishness story stopped me in my tracks—it paralyzed me. I have so many ideas of ways I can help my corner of the world, and I'd like to try them. But what would be the point? If everyone is selfish, my efforts are just a drop in the ocean and trying to help would be futile, so I might as well not try.

Fourth, and this is the most insidious part, the selfishness story gave me a certain kind of complacent comfort. Because if it is true that people are selfish, then I get the luxury of feeling superior without actually having to do anything (which, according to the story, would be pointless).

If people are selfish, I don't have to confront other people's complacency by the actions I take. I don't have to try to navigate the confusion of not knowing what to do. I can continue to find reasons to not pick up the challenge of being fully myself. I have an excuse not to grow and change.

Next question: *Who would I be without that story?*

In my own case, I enjoy being playful and friendly, but I hold myself back when I feel suspicious of other people and their motivations. I find that when I don't cling to the story that people are selfish, I'm more open and gregarious, joking around and having more fun without all the fear.

Also, when I challenge the story, I am more able to trust that the people and the world around me are generous enough for me to get the support I need to live and thrive. That leaves me freer to be generous as well, especially with my time and effort. I feel safer dedicating myself to working on the issues in the world that I care about.

Of course, I am still sometimes scared to believe that I can make a difference with my actions. But as I practice using my power, I get more and more used to it. In short, without the humanity-is-flawed story, I feel freer to be myself. It's much better for me and better for the world if I don't hold on to this story.

TURNING THE STORY AROUND

The last stage of Byron Katie's work on stories is to turn them around to their opposites to develop alternative thoughts that may bring you peace and help you function better. There are lots of ways to turn a story around. You can turn a story about others around on yourself. "People are selfish" becomes "I am selfish." You can turn it around on the object: "People are generous." And you can turn it around to its opposite: "People actually should be selfish."

The point of the turnaround is to find an alternative view that helps you better deal with reality. When it comes to the selfishness story, here is the turnaround that I most like to work with:

"Sometimes people, including myself, act out of selfishness, but all things being equal, most of us prefer to live out of generosity and to help each other when we can, and there is every reason to believe that it is worth trying to harness that goodwill to help the world."

What would happen if I acted as though this turnaround were true? Well, for one thing, instead of dismissing humanity's character, I might wonder what conditions bring out humanity's best and try to create those conditions in my work and personal life. Suddenly, instead of being paralyzed by the cultural story about the enemy within, I can see a path forward.

The point is not to replace the old story with the new story. It is just to release our grasp on the old story enough so that our natural love, compassion, power, and freedom can come out.

WHAT HAPPENS WHEN YOU LET GO OF THE SELFISHNESS STORY

In 2000, the founders of ice cream company Ben & Jerry's Homemade, Ben Cohen and Jerry Greenfield, tried to stop the international conglomerate Unilever from buying out their company. They worried their company's social consciousness strategy would be sidelined. They assembled a group of financiers to offer a competitive bid, but when the group couldn't offer as much as Unilever, Ben and Jerry's investors sued to force through the Unilever purchase.

People are selfish, right? They care only about money and their own pocketbooks. There is no point trying to do good in the world because selfish people will always run you over, right?

That's not what a group of three friends who went to Stanford University together believed. Knowing what happened to Ben & Jerry's, the Stanford grads have paved the way toward a new corporate structure—known as the "benefit corporation" or B Corp—that allows company founders and officers to maintain their companies' social missions even if it means investors make less money.

Instead of buying into the "people are selfish" story, B Lab's founders—Jay Coen Gilbert, Bart Houlahan, and Andrew Kassoy—assume human nature is *not* selfish. Jay told me, "There are very few human beings who don't want to leave the world a better place than they found it. People want a purpose. For most people, meaning is more important than money and purpose is more important than profit."

It turns out that the problem that has long dogged socially responsible businesses is that public companies are required by law to maximize profits for shareholders no matter what. Inviting investment in a socially responsible business in order to grow it, therefore, has always opened the company up to the risk of losing control of its mission. The problem is not selfishness but institutional structure.

"What we needed to do," Jay said, "was to make it easier for businesspeople to do what they wanted, which was to bring their whole selves to work, and also to make it easier for socially minded investors to find the kind of businesses they want to support."

So Jay and his fellow cofounders created B Lab in 2006 to certify companies that amend their articles of incorporation to say that managers must consider the interests of employees, the community, and the environment instead of worrying solely about shareholders' profits. Those amendments let entrepreneurs take on outside investors without worrying that their values will be compromised.

"This allows founders to sell their company to bidders who care about the mission. Investors can't sue you and force you to sell to the highest bidder."

By virtue of having their companies classified as B Corporations, the founders of both Plum Organics and Method Products, for example, were able to sell their companies while ensuring their social mission would stay in place. "In the past you had to stay small or sell out," Jay said. "These sales made it clear that there was a third way."

As of this writing, there are 910 certified B Corporations in sixty industries, with familiar company names such as Seventh Generation, Etsy, Cabot Creamery, Patagonia, King Arthur Flour, Rubicon Global, and many more numbered among them.

And all because the B Lab founders let go of the "people are selfish" story.

Paralyzing Story #2: The Unbeatable Enemy Without

Government/corporations/the wealthy/the political parties are too strong. There is nothing anyone can do to change them, and if there is going to be change they have to initiate it, so all of our individual efforts are irrelevant.

This second cluster of stories I hear over and over again says that no matter how good human nature may be, elements of human society—government, corporations, and the like—represent an out-of-control force that can't be overcome. It is their job to fix things. Even if they don't, you can't beat the system, so you'd best just keep your head down and go along to get along.

If that's true, how could there be so much change in the world in spite of attempts to maintain the status quo by government and big business: gender equality, marriage equality, improvements in world health and life expectancy, and so on?

On a more granular level, how much influence do governments and corporations really wield in our individual corners of the world? Can concentrations of wealth stop every effort we might make to improve our lives and the lives of others in our communities?

So, is it Absolutely True that some enemy without makes our individual efforts irrelevant?

No!

Once again, the simple realization that we are holding on to a story at all is enough to make us see that we are limiting ourselves: "Oh, I see why I feel so stuck about starting a community garden! I've been telling myself a story about how no matter what I do, the agricultural corporations like Monsanto and factory farms will control our food system forever. But if that's true, why are there so many people shopping at our local farmers' market every Saturday?"

How do I react, what happens, when I believe this story about big societal forces making our individual efforts unimportant?

The idea that any societal force is so large as to make my efforts pointless makes me feel small and meaningless. Plus, it denies the many good things I have done and seen other people do. It makes me feel doomed. How does it make *you* feel?

Who would I be without that story?

Well, I wrote a whole book about this. My previous book, *No Impact Man,* is the story of what happened when I decided I would stop calling myself a victim of societal forces. In short, my life as I knew it unraveled and I became free to be the person I am now: to try to help the world when I want to, to have fun when I want to, and to have fun trying to help the world when I want to. Who would *you* be without this story or any of the other stories that stop you?

TURNING THE STORY AROUND

How about this?

"Government/corporations/the wealthy/the political parties are too ~~strong~~ weak. There is nothing ~~anyone~~ they can do to change ~~them~~ themselves, and if there is going to be change, ~~they have to initiate it~~ we have to help them, so all of our individual efforts are ~~irrelevant~~ crucial."

HOW TO CHANGE A CITY BY NOT BELIEVING THE "GOVERNMENT IS TOO STRONG" STORY

In the early 2000s, Kate Zidar was one of many New York City citizens who refused to wait for the city government to institute a composting program to manage food waste. Instead, she started her own community compost pile in a corner of a city park. Compost piles like hers popped up in other communities all around the city. In 2013, seeing the success of these compost piles, Mayor Michael Bloomberg's government finally announced that it would move toward a citywide curbside compost program.

Here is what Kate says:

> *Back in the early 2000s, I was volunteering in a community garden in Williamsburg, in Brooklyn's McCarren Park, and my focus became soil toxicity. There was a lot of dioxin*

and lead in the soil because of fallout over the years from a nearby incinerator. I wanted to replace the top layer of soil so we could grow food safely.

Also, trucks carrying food waste and other trash in the solid waste management system screwed up the air quality in the area. So it made sense to both create good topsoil and divert food scraps from the waste stream by starting a community compost pile.

In McCarren Park, there was a part of a dog run that was not used, so I "annexed" it. I wrote a letter to the parks commissioner including a map showing the location of my new compost pile—I also sent a flower bulb, hoping that would get his attention—and asked for permission. He never wrote back, but I kept a copy of my letter and I told anyone who tried to interfere that the parks commissioner knew about the compost pile.

In addition, I used really heavy fifty-five-gallon plastic drums to house the compost system. They could not be moved easily. My idea was to make it so the work involved in shutting the compost system down would be greater than whatever problem park workers seemed to feel it caused.

At first it was just me hauling my kitchen scraps to the barrels. But passing foot traffic soon attracted random people dropping off their food scraps, too. Before long, a woman named Jo Micek started to help. She was a community organizer and she knew how to raise funds. Pretty quickly, the compost pile was being run by a "dirty dozen." (Get it?)

Not long after that, there were more than one hundred families dropping off their food scraps every week and the compost project turned into a collective, not just run by me. Meanwhile, the compost went back into the community garden, home gardeners took it home, and eventually even the park workers began to use it around the park.

Why didn't we begin by going to the city government and asking them to start a compost pile for us?

Everyone who works in community gardens knows that the gardens start by essentially squatting on an abandoned, unused piece of land. You don't start by working with the government but by working with your community for improvements everyone wants. When you try to work with the city agencies, they stonewall the idea because they have a whole range of missions and obligations to consider, but you have only one—your garden or compost pile.

I didn't want to use my energy dealing with the bureaucracy. I wanted to compost. Plus, I knew the project would actually represent a community improvement. I didn't want to ask for permission. I could always later ask for forgiveness. Ultimately, there was no way the parks department could stop it because it became so popular with the local community.

This is one way to bring about broader city or social change. You don't ask the government to do it. Instead, you gather with other citizens and you demonstrate to the government that it is needed, is wanted, and works. That is why New York is adopting curbside composting now. Because so many communities like ours demonstrated that composting is needed, is wanted, and works.

Meanwhile, the personal benefits to me were the people I met and the friends I made. Also, I figured it out on my own. I didn't know what I was doing and I started it and saw it through. I developed my own system of doing something. Once you do that in one area of your life, then you can do it in all areas. It made me less uncomfortable with not knowing how to start.

Paralyzing Story #3: Utopia Is Coming!

Technology/God/evolving consciousness/the planet's systems/the government/the free market/the perfect Oneness of the universe will save us. Since that power for good will take care of things, all of our individual efforts are not just irrelevant, but unnecessary.

The utopia story is particularly seductive because, unlike the other two, it is often married to a feeling of superiority over the "pessimists."

Of course, it is true that there are forces for good that can help with the world's problems. Not everything is going downhill. That's the good news. But the perfection of the universe (for those with a Buddhist or New Age bent) or of God (for those who prefer Abrahamic approaches) includes the compassion and desire to take action that arises in our hearts when we see suffering.

To deny our individual agency is to refuse to be the hand of God or the universe or any other force for good. There is no movement toward utopia (even if that is what you hope for) without our participation.

If it is technology we put our faith in, will the technology design or deploy itself without us? Will it take care of all aspects of the world's troubles, or might we have to complement the technology with some sort of human change? In the case of climate change, say, might we also have to stop burning fossil fuels?

Is it Absolutely True—can we say with complete certainty—that technology, God, evolving human consciousness, or any other force for good will solve the world's problems without the need for the efforts of individual humans—in short, for me and you?

No.

How do I react, what happens, when I believe this story?

As I write, a story is going around the Internet that a new kind of boat will be able to roam the oceans and suck all the plastic out of them like a gigantic aquatic vacuum cleaner. On people's Facebook walls, I see the link to this article with status updates like "I am so relieved." What the status updates don't say is: "And now I no longer have to worry about this."

That is the effect of the utopia story—whether it is technological utopia or spiritual utopia or governmental utopia. We use the story to make us less worried so we can stop paying attention. But the thing is, if I am not here to pay attention and participate in and help at least my corner of the world, then what, actually, is my purpose?

If we hide in our philosophical caves until life is over, then life will be, well, over.

The utopia story makes me feel meaningless, as if I don't have a purpose when it comes to making a better world. It also makes me feel stressed. Because the reaction of my True Self is to act and to take initiative. But this story tells me I should rein that in. Worse, if I believe utopia is coming when it is not, then I end up ignoring a big world problem that is actually still getting worse and desperately needs our help.

Who would I be without that story?

The good things I see in government/technology/the universe/etc., instead of making me ignore my ability to contribute, would energize me and inspire me to believe that I, too, can make a difference.

TURNING THE STORY AROUND

Here is the version I like, because it offers me some of the comforts of the original story without turning me off to the world:

"In some ways, maybe utopia is already here. After all, we have technology/God/evolving consciousness/the planet's systems/the government/the perfect Oneness of the universe to help us with our problems. Since that power for good will help us take care of things, all of our individual efforts are so much more likely to be helpful. Everything is not perfect. But maybe the only perfection we can really ever find is how we allow ourselves to respond to the imperfection. Maybe that is utopia."

WHY BELIEVING THE "UTOPIA IS COMING" STORY MAKES US SAD

> *Suppose there were an experience machine that would give you any experience you desired. Superduper neuropsychologists could stimulate your brain so that you would think and feel you were writing a great novel, or making a friend, or reading an interesting book. All the time, you would be floating in a tank, with electrodes attached to your brain. Should you plug into this machine for life, preprogramming your life experiences? . . . Of course, while in the tank you won't know that you're there; you'll think that it's all actually happening. . . . Would you plug in?*

Harvard professor and author Robert Nozick first posed this thought experiment, known as the Experience Machine, in his book *Anarchy, State, and Utopia.* It is one of the most famous thought experiments in contemporary philosophy. The overwhelming majority of people, Nozick suggests, would not plug into the machine.

The reason, he says, is that we want to actually *do* certain things, not just have the experience of having done them. We want to *be* certain people, not just get the results that would come if we were those people. "Perhaps what we desire is to live ourselves," Nozick wrote, "in contact with reality."

Most of us who perceive things wrong in the world don't just want them fixed. We want to be able to say we were part of fixing them. The idea that there is some utopian force out there that takes away the world's needs for each of us as individuals suggests a meaningless existence where, yes, we can feel pleasure, but, no, we don't matter.

Most people don't actually want to live in that world. A force for good, most of us feel, is excellent, but a force for good that requires and needs our participation is even better.

An Exercise in Getting Rid of the Stories That Block Your Lifequest

1. Take a moment to dream. Be completely what our parents, teachers, and bosses call "unrealistic." Override your inner objections and indulge in complete fantasy. Ask yourself: How would the world look if it could look whatever way I wanted? What part would I play in bringing that world about? How would I want my life to be improved as I played that role in terms of security, meaning, and fun? (Remember: fantasize.)

2. Now, make a quick list of the reasons you believe these things can't happen—we're back, again, to those stories we tell ourselves. Apply the following questions, slightly adapted from Byron Katie's, to each of your stories. You can apply them to any story that stands in the way of you and your search for the life choices that might make you and the world happier:

a. Is this Absolutely True?

b. Even if you believe it is true, can you know for sure that is true?

c. How do you react when you believe this story?

d. Who would you be without this story?

e. How can you turn the story around in a way that doesn't divorce you from your Truth and capacities?

3. Ask yourself, what would your small corner of the world look like if your vision for the whole world were true? For example, if, in your vision, we all pitched in with taking care of children, how would that look in your home, on your street, and in your neighborhood?

4. Think of one small action you could perform without hardship in the next twenty-four hours that could move your home, street, or neighborhood toward that vision. Perhaps, in the example of our caring for the kids together, you could stop and chat with the first child you saw. Come up with your action while letting go of your perfectionism and the idea that it would be pointless if there was no follow-up.

5. Within the next twenty-four hours, do it. Just that one action. Don't question it. Just do it.

6. After you have done it, if you liked it, think of one other small action that moves your life and your world toward your vision, and commit to doing that.

7. Now, don't ask yourself if your action helped the world. Instead, ask: What was the effect on me of taking that action? Did it make me happier?

2

UNDERSTANDING THE TRUTH OF YOUR RELATIONSHIP TO THE WORLD

Your life = All your relationships added together
All your relationships added together = Your relationship to the world
Therefore, your life = Your relationship to the world
which means
a. What you do in your life affects the world
b. What happens in the world affects your life
c. You are important to the world and the world is important to you

It's In Front of Your Nose and Under Your Feet

"Better," Lara said when I asked her how she was.

"Better than what?"

"Better now that my parents have gotten in touch. For eight days after the typhoon hit, I didn't know where they were or if they were alive."

It was November 2013. Typhoon Haiyan, the strongest storm ever recorded when it hit land, had just killed thousands in the Philippines. I had been so sad about what I thought had happened so far away. But seeing my friend looking drawn and exhausted, I realized that Haiyan happened not only far away but also right here in Brooklyn. It didn't happen only in the Philippines. It happened in the world as a whole. In Lara's world. And therefore, in my world.

Sometimes we get so caught up in our own lives that the world beyond them feels impossibly distant and removed. But that world—*the* world—is right in front of our noses and right under our feet and we are all more connected than we realize. Your family is the world, your food is the world, your trash is the world, your friends are the world, your car is the world, and on and on. Every aspect of our lives connects to every other part of the world in ever-expanding spirals.

There are some people who believe that we are also all connected mystically, like trees that appear separate above ground but share a single root system underneath. If one of those trees is injured, the whole organism is injured. And it's the same for us. Maybe we are connected mystically and maybe we are not. (Personally, I say yes, but who knows for sure?) Regardless, seeing Lara's tired face reminded me that we are certainly connected practically. You connect to the entire world through all of your day-to-day relationships.

So:

Your relationship to the world = your friendships + the stuff you buy + your work + how you spend your money + where you live + how you form your family + your participation in civic life + the way you treat taxi drivers + how you travel + your relationship to corporations + the air you breathe + how much fun you have + the way you suffer + the way you treat other people who suffer + the amount you sleep + the coffee you drink + the food you buy + all the other little relationships that fill your life.

What important conclusion comes from this? When you change any one of these little relationships, you change your relationship to the world. When you change your relationship to the world, you change your life. *And you change—in ways large and small—the actual world.*

It gets everything you give and gives everything you get.

Trust me for a minute while I say this: to have a good, happy, fulfilling, fun, sexy life, you really need only two things—security and meaning.

> Security is what you need to be alive, safe, healthy, and comfortable, and comes from *what you get from the world through your relationships.*
>
> Meaning is the buzz you get from being alive, and includes self-expression, the ability to change things, adventure, and service, all of which come from *what you give to the world through your relationships.*

We will delve more deeply into what it means to be secure and what it means to have meaning later on. But for now bear with me. What I am saying is that: (1) a good life requires both getting and giving; (2) everything you get comes from the world and everything you give goes to the world; and (3) all that giving and getting with the world happens through your many relationships.

Let's look at how this principle applies in the case of one relationship: food shopping. Going to the grocery store sucks, right? The lighting is awful. The freezers make it cold. Plus, we just can't feel good about squeaking our carts through the aisles knowing that much of the food is filled with chemicals, harvested by poor workers, and the result of intensely cruel treatment of animals. You spend half your time worrying about getting food that is actually good for you, and the rest trying not to get ripped off.

What happens at the grocery store kind of sucks for the world and sucks for you. Many of us just can't wait for the experience to be over.

That's what so many of our relationships to the world are like—we just want to get them over with. We feel stuck with doing things we totally hate—the ride back and forth to our jobs in terrible traffic jams is another example—to meaninglessly eke from the world the things we need. How many of our relationships are like that? How much of our lives do we just *want to get over with*?

The problem is that we've come to expect so little of our relationships with our worlds. Doing what you hate to get what you need is a sign of ridiculously low aspirations. We spend hours every week shopping for food, for example. Why should we accept this when it sucks?

We can do better. The most fulfilling relationships with the world meet the needs for both security and meaning all in one place. A relationship through which you both give and get—one that provides both some security and some meaning—is a relationship that kicks ass.

Why shouldn't we have relationships like that in every area of our lives? In the case of our food, for example, patrons of farmers' markets tend to love their shopping. In addition to getting food that is safe and nutritious, they get to support a food system that includes farmers who actually care for the land and for the well-being of their livestock. People who shop in farmers' markets both get and give—security and meaning.

Not everyone has access to farmers' markets, of course, but the point is that the most fulfilling relationships to the world include both security (getting) and meaning (giving).

The Principle of Barakah

In Islam, barakah is a divine force that can be present in objects, people, transactions, and relationships in the material world. It brings prosperity, protection, and happiness. It embodies the property of being blessed and bringing blessing to everyone involved.

In a business transaction or a relationship, for example, the divine force of barakah will only be present in a transaction or relationship that does good and is kind to all involved. Meat coming from an animal treated cruelly will not carry barakah to the eater. Oil extracted from land won by war cannot carry barakah.

The principle of barakah is that when we work to ensure that our transactions are good for everyone involved, those transactions will ultimately be better for us. Barakah is present and we will be most blessed by a relationship and transaction only when all parties in the relationship give and receive kindness.

Whether or not you believe in divinity or blessings, isn't that the kind of relationship you'd prefer to participate in? The kind that adheres to the principles of barakah? The kind where you both get and give both security and meaning?

IT FIXES YOU WHEN YOU FIX ANY PART OF IT

My friend John worked his rear end off as a magazine editor and volunteered in his free time in a homeless shelter. He got his security from working and his meaning from volunteering. Doing both, he had no time for himself. Finally, he went back to school and got a job as a social worker. Suddenly, he had a lot more free time because he got his needs for meaning and for security met in the same relationship.

Part of why we are all so stressed is because our lives are filled with half relationships that don't fill our needs to both get and give. So it takes us more time and more energy to have what we need because we have to service too many half relationships. Not all of us are in the position to change our jobs or want to change our jobs. But as important as the standard life approach tells us our jobs should be, they are not everything, nor are they how we do everything.

The way to begin to fix your relationship with the world is to choose just one of your relationships and fix it. Think of what happened when my friend Kate changed her relationship to food waste.

Your life improves if any one of your relationships improves. Similarly the life of the whole world improves if any one of its relationships with one of us improves.

So if you improve your relationship to the world, you make your life better and the world better. Here is a central principle of this book: *The more relationships you have that reflect your true values, the happier you will be. The more relationships you have that reflect your true values, the happier the world will be.* One relationship at a time, you fix your life and fix the world. How to fix each of your relationships is a lot of what the rest of this book is about.

How Do You Relate?

Don't worry about the world. Don't worry about your life. Instead, think about the tiny little relationships that make up your life and the way you relate to the world. One at a time, you can change those relationships, and one day you will wake up and find yourself living a better life in a better personal world.

Now, don't make hard work of this. Just take ten minutes and use an old scrap of paper or type on your phone. Make a list of your relationships. Not in any particular order. In fact, the order in which they come to you might in itself reveal something important to you.

Write down the relationships through which you get your food. Your clothing. Your shelter. Your money. Your friendships. Your sex. Your meaning. Also the relationships through which you assert influence on government and institutions. The relationships through which you give what you want to give and any others that come to mind.

Now look at this list and ask yourself what about each of those relationships gives you security and what gives you meaning? Which of the relationships gives you both? Which gives you neither? Are there any half relationships you could eliminate to replace with full relationships? Are there any relationships you could combine?

Choose one relationship that sticks out for improvement. What is the easy, immediate thing you can do today? If you did one easy thing about that relationship each day for the next week, what would those easy things be? Do them.

3

THE LIFEQUESTER'S MIND HACKS

Some are perfectly satisfied with what they have; they eat, drink, have children, and take life as it comes. Others can never forget that they are being cheated; that life tempts them to struggle by offering them the essence of sex, of beauty, of success; and that she always seems to pay in counterfeit money.

—COLIN WILSON, INTRODUCTION TO *THE OUTSIDER*

Suppose you admit that you feel psychologically or physically or practically or spiritually or emotionally incapable of conforming to some or all of the standard life approaches and their associated stories and half relationships. Suppose you find the priorities of our society and its ways of doing things at odds with your values and perspective. Suppose you want to relate differently. Suppose you can't fit in. Or you simply don't want to fit in.

Without stories and standard approaches and ways of relating you can believe in, what are you supposed to do? How are you supposed to live?

Back in the early 1950s, these were the questions that plagued Colin Wilson, the young son of a shoe factory worker from Leicester, and bestselling author of *The Outsider* and many other books. Wilson dropped out of school at age sixteen because he couldn't understand the relevance of what he was being taught to real life. He also quit his job in a wool warehouse because he found it meaningless as well. He got called up for British military service and managed to get dismissed after six months in the Royal Air Force by falsely claiming to be gay.

He felt incapable of going along to get along. He just couldn't make himself follow the normal path. No one around him seemed to understand. His father nearly disowned him. Wilson felt there was something deeply and terribly wrong with him.

After failing to fit in at yet another job, this time as a lab assistant, Wilson was so distraught that he contemplated swallowing the contents of a vial marked *hydrocyanic acid* (commonly called cyanide). As he turned the vial over in his hand and thought about removing the cap, he had a sudden insight. Within himself he recognized that there were actually two Colin Wilsons, with two differing attitudes toward the world.

One attitude belonged to the persona of, in his words, an "idiotic teenager" who felt that drinking cyanide was a solution. In this persona, Wilson felt nothing but self-pity over the fact that he could not fit in. But the other attitude Wilson recognized within himself belonged to what he now realized was his True Self, the part that "glimpsed the marvelous, immense richness of reality, extending to distant horizons."

Over time, from this perspective, he began to suspect that the reason he didn't fit into society was not because of a problem with him. The problem, he began to think, was with society. He wasn't the "idiotic teenager" he had so harshly judged himself to be. He simply was less able to ignore the feeling of alienation—of being lost—that we all increasingly suffer from. Perhaps, in some ways, he was just less able to ignore that society actually was *absurd*—in the full sense of Albert Camus's meaning when he used the word to describe the gap between humanity's desire for meaning and its inability to act meaningfully.

Perhaps Wilson had a point.

A voracious reader, Wilson identified with creative misfits in literature and the legions of writers and artists who portrayed them, from Camus and Ernest Hemingway to William James and Vincent van Gogh. Wilson, who by now had made himself a self-taught expert in existential literature, went to work on a book examining the role of social outsiders like himself in classic works of art and literature, writing all day in the halls of the British Museum. To save money so he could afford to write, he slept in a sleeping bag in Hampstead Heath.

At the age of twenty-four, in 1956, he published *The Outsider*. It was his first book and it became an instant best seller, going out of stock in stores around the country on its first day. *The Outsider* explored ideas of social alienation and portrayed the meaninglessness of the mainstream and its standard life approaches. Wilson wrote about how anyone who perceived this meaninglessness faced the terrible choice between alienation from himself and alienation from society. Outsiders could fit with society or fit with themselves.

At first, the literary establishment embraced Wilson as a brilliant young writer. But they had made the mistake of thinking that by calling himself an outsider, Wilson meant that a working-class man like himself did not have access to standard success and privilege. In fact, his writing pointed out not that he couldn't become privileged but that he despised the values of the privileged. That the chase for standard success offered no more than booby prizes.

In the ensuing flurry of press interviews, Wilson repeatedly and publicly made clear his disdain for elitist values. When his future father-in-law discovered notes for a novel about a sex murder and assumed they were his journal, the literary establishment took its chance to punish Wilson for his contempt. Where they had once heralded him as the founder of a generation of "angry young men"—the British literary movement of disenchanted writers—they now denounced him in the newspapers as a sociopath. Wilson ran away from London to the countryside with his fiancée and proceeded to spite the literary establishment, repeatedly establishing his genius by prolifically writing and publishing nearly two hundred books throughout his career.

Sixty years later, *The Outsider* remains in print and has sold more than a million copies. Over time, Wilson, who died in 2013, developed a huge cult following. And here's the point: Wilson's popularity grew because more and more of us began to feel the same kind of alienation that he experienced and realized for ourselves what he had been saying all along: that many of the priorities of society are absurd.

Back in 1956, *The Outsider* spoke largely to creative types who could not fit into the cookie-cutter molds of business and industry. But since then, "society" and "culture" have drifted even further away from the humanity they are supposed to support. It's no longer just the artists who feel they can't fit in.

Maybe you can't sit for hours in a cubicle without getting a diagnosis of ADHD, or you're gay or a person of color, or the nuclear family structure doesn't suit you, or you're introverted in an extroverted society, or you're a boy in a public school who can't wait until recess, or you're differently abled, or you don't identify by gender, or you're a single parent, or you don't want to be a parent at all, or you don't care about money and stuff, or perhaps you simply aren't interested in going over the planetary cliff with the rest of the mainstream.

As the mainstream suits less and less of humanity, and as the crises mount and mount, more and more of us feel like misfits, outsiders. The question is, do we have the inner strength to conclude, as Wilson did, that the problem is not with us but with the mainstream? Do we have the strength to trust ourselves instead?

Of course, having come this far in your thinking, you could take a dark turn. You could choose to save only yourself. You could, for example, pretend to fit in and push even harder at society's door by working harder or getting more qualifications or buying a bigger house. Or you could disavow society and materialism and become a spiritual recluse. You could disappear into the mountains as a survivalist and hope to outlast society's self-destruction. The thing is, these, too, have become standard life approaches.

These standard approaches to our alienation, these various ways of withdrawing from society, Wilson once said to an interviewer, are like those of people who wish for the Land of Oz. They have, he said, "this desire for the place over the rainbow, and [they feel] once [they've] glimpsed it that life isn't worth living."

Wilson, on the other hand, felt optimistic about human society. He didn't think it was something to run from. Instead, he thought it was endlessly changing and growing. Our view of society should not be how far it had to go but how far it had come, especially in the last two hundred years, when so many humans have moved from mud huts to skyscrapers.

Wilson wouldn't think the one billion who don't have access to clean drinking water proved the failure of society, for example. Instead, we should celebrate that six billion people *do* have clean drinking water, and consider that fact proof that society is actually capable of getting good water to the rest. We should not pull our hair out over climate change but should celebrate the international movement that seeks to do something about it—and join in.

Society, Wilson believed, has come a long way. The job of the outsider is not to run away but to stimulate it to go even further. Wilson told another interviewer that outsiders have to stop "saying you don't want anything to do with this lousy material world, because if you don't do something about this lousy material world, nobody else will. You yourselves have got to take over and become the leaders."

In other words, the fresh choice is to become what we have been calling a lifequester. To find a path back to ourselves and what we value and then to show that path to others. A lifequester is someone who breaks away from the cultural stories to discover who she really is and what makes her truly happy and fulfilled and uses those discoveries to help the world.

The lifequester is a sort of bushwhacker. We slice through the branches for ourselves but in doing so clear a path for others. As we help each other find a path back to our Selves, we help society find a path back to alignment with the human values required to live compassionately and kindly in a planetary habitat with limited resources.

Lifequesters take it as an article of faith that when you let go of the standard life approaches and the societal stories, you are not adrift but are finally in the company of the one thing that you need to carry you and the world home—your True Self. According to Wilson, when you find that you are alienated from society's stories and standard life approaches, you are not supposed to leave society. You are supposed to lead it.

We are used to the idea that we can influence our own lives. What we don't talk about so much is how the meaningful part of life is when we influence the world—from the lives of our loved ones to those far away. To many, happiness is the idea of figuring out who you are and what your influence is. What stops us from trying to help the world is that we have grown unaccustomed to the idea that we influence the world. That muscle has atrophied.

So, before we move on, we are going to discuss a few basic concepts that many lifequesters I've spoken to take to heart. These are concepts that make lifequesters feel optimistic and give them energy to move forward. They focus on the nature of our relationship to the world. On why you will be more productive and help more people if you look more for the good in the world than the bad. On the mechanisms by which each of us can make a difference. On choosing to work in the areas of life that you actually control. On how our well-being and the world's well-being are connected. On how there is no such thing as self-help without each-other help.

These are some things to keep in mind as you read this book.

Look Away from the Obstacles

When you're driving a racecar, you go about two hundred miles per hour. You have to wiz around a lot of crazy corners and there are a bunch of other cars weaving in and out and all around you. There are lots of obstacles you need to avoid. You could easily hurt yourself or someone else. Everything feels out of control.

In other words, to my way of thinking, it's a bit like being alive in the world right now.

Your body starts feeling pretty tight and anxious. Your thoughts, if you're anything like me, keep going to the dangers. Which is why, when we are discussing the search for a better life that contributes to a better world, you might expect a discussion about the problems.

In the climate movement, in the social justice movement, in the food movement, in all the change movements, we are fond of repeating the bad news. People are starving. The planet is warming. The economy is collapsing. Our old routes to personal happiness no longer work.

These stories may have been excellent for getting you to wake up and pay attention. Perhaps they made you decide to be a lifequester or to buy this book. But once you're ready to take action, they start to feel overwhelming. They take away your energy to act and to imagine.

What a racecar driver will tell you is that if everything feels crazy on the track, if the car feels like it is about to go into a spin, you shouldn't stare at the obstacles you might crash into. It isn't the obstacles you need to be most conscious of.

The last thing you see before you crash is the thing you hit. It turns out that your hands unconsciously steer toward what you're looking at. To avoid a crash, you need to not look at the obstacle. You need to look at the clear road where you want your car to go.

That's what lifequesters do. They don't waste time pulling out their hair over what is wrong with the world. They don't worry too much about how our lives aren't as happy as we want them to be. Instead, they concentrate on the spaces. On their vision. And then they look for the clear road that leads to where they want their lives and the world to go.

Because, according to the racecar drivers, we steer where we look. The trick is not to look at what you're scared of crashing into. The real trick, it turns out, is to look toward where you want to steer.

Don't Waste Your Energy on Finger Pointing

A sober friend, Jesse, was once trying to explain to me the twelve-step principle of "taking your own inventory." He told me about a time when he was in terrible conflict with his girlfriend, Molly. Every day they argued and every day he phoned his twelve-step mentor to complain about her.

Jesse would explain to his mentor in careful detail all the things Molly said and did wrong. Invariably, his mentor would ignore the story, latch onto something Jesse said or did, and then suggest an improvement to Jesse's behavior. The mentor completely ignored Jesse's complaints about Molly.

Every day, my friend got angrier and angrier with his mentor. Finally, after a month or so, he blew up. He said to his mentor, "I keep telling you all the mean things Molly says and you just ignore them and tell me ways to act better. Can't you hear how much she's hurting me?"

"Yes, I can," his mentor said, "but our job is to help you look for your part in the problems."

"But why? It isn't mostly my fault. It's mostly her fault. What is the point of looking for my part?"

There was a pause. Then the mentor said, "Whoever's fault it mostly is, yours is the only part we can actually change."

Molly had done some awful things. But the thing is, no matter how hard Jesse wished for it, talking to his mentor couldn't change her. If they looked carefully, his mentor said, "We may be able to improve the situation a little by working on the one person we have control over—you."

When you look at the world—the hunger, the corruption, the injustice, the poverty, the inequality—it is so easy to blame the senators or the CEOs or the bankers or all the people who don't seem to care. It is so easy to spend so much time and energy thinking of all the ways they are wrong. And it may very well be that the problems aren't mostly our fault. The problems are mostly their fault.

But as long as we waste our energy looking to put the blame elsewhere, we put the power elsewhere, too.

"Taking your own inventory," it turns out, is like a magic trick for discovering where your power lies in any difficult situation. When a situation causes you discomfort, you put aside whatever blame you feel toward other people or institutions and write down the ways you contribute to the problem.

If you are worried about the lack of citizen participation in our democracy, you might write down the ways in which you don't participate or time you don't spend trying to help others participate. If you care about world hunger, you might write down the things you could do to help that you aren't doing.

No matter how much you blame corporate money for corrupting our democracy or big banks for causing developing-world debt and hunger, figuring out how you participate in the problems provides you with a route to working on the solutions. Do you march in the climate protests? Do you speak to your elected representatives? Do you build the kind of community you want? Are you working on your vision for your life and the world?

This is not about self-blame. In fact, concerning yourself with the beam in your own eye instead of the speck in your neighbor's is not a pious attempt at self-flagellation.

Self-inventory, when executed with love and compassion toward yourself, is actually a miraculous tool. Because finding how your life intersects with the problems provides you with your leverage and points to solutions. Those are the bits you have power over. That's where you can start your work. This is how you begin to change your energy from blaming to fixing. From being a victim of the universe to being a co-creator of it.

Limit Your Concern to Grow Your Influence

During talks I've given around the world, I've met a lot of resistance when I have suggested the use of self-inventory to look for our own parts in global problems and to start by addressing those first. People have objected: *I am such a tiny cog in such a gigantic machine. What is the point of my doing anything?* The thing is, if you worry about the gigantic machine instead of fixing the small cog, nothing at all gets done. When your concern ranges beyond the things you can control, your influence shrinks to zero.

Consider this parable:

One day, a man went to the beach and saw that a storm had washed thousands of starfish onto the sand. They baked in the sun and began to die. The man wished he could help, but he became overwhelmed by the scale of the problem. How could he possibly save all these thousands of starfish? With a sad heart, he turned to leave.

Just then, a little girl arrived. She took in the devastating scene and walked toward the nearest dying starfish. She picked it up, carried it to the water, and dropped it in. Then she picked up the next nearest starfish, and the next, and the next. The man watched her for a moment, thinking about how trying to save all the starfish was futile.

Finally, he walked up to her. He said, "Little girl, there are thousands of starfish dying on this beach. You can't possibly make a difference."

She paused for a minute, looked confused, and then said, "Tell that to the one I just threw in the water." She carried on with her work. After a moment, the man joined in. As the day progressed, more and more people arrived at the beach and, seeing what the little girl and the man were doing, they joined in, too. By the time the day was over, many hundreds of the starfish were saved.

In the beginning of the story, the man allowed his area of concern to stretch beyond his area of influence and thus paralyzed himself. On the other hand, the girl restricted her area of concern to her area of influence, and her area of influence actually grew because her efforts encouraged so many others.

In real life, in 1997, my friend Julia Butterfly Hill climbed into the branches of a 180-foot-tall, 1,500-year-old redwood tree in Northern California and began living on a platform at its top to stop loggers from chopping it down. Julia ended up living in the tree for 738 days—more than two years. By the time she came down, the loggers had felled all the trees around her. She was able to secure an agreement from the logging company to save only the one tree and those within 200 feet of it.

In the years since, Julia has spoken to crowds totaling in the hundreds of thousands. She has given thousands of press interviews, has had books written and movies made about her, and has inspired many thousands of people to join the environmental movement. Like the little girl with the starfish, by limiting her area of concern to the one tree she could help, Julia has inspired others to save millions of trees.

Major P.S.: Julia has also made a really good living and life for herself doing this work.

Be Too Stupid to Know Your Limitations and Take the First Step

When I launched my No Impact year, my first step was just to begin living with the lowest possible environmental impact. A few people said to me that I was "too stupid to know that one person can't make a difference." Being too stupid to know your limitations, however, can be an amazing asset. Think on this story about two frogs:

The two frogs—one very smart and one very stupid—are caught in a bowl of cream. The sides are too steep to climb and the frogs have no foothold to jump. The stupid one begins to swim as hard and fast as he can. The smart one looks over and says to himself, "He's too stupid to know that all that effort will make no difference."

Having weighed the hopelessness of the situation, the smart one decides that the most intelligent thing is to give up. So—*blub!*—he drowns. The stupid one keeps trying. Just when his legs are about to give out, the cream starts to get thicker. His struggling has churned the cream to butter. He's surprised to find himself on solid ground. He jumps out. Because he stupidly pursued the first step (swimming), the second step (jumping out) appeared, as if by magic.

The question is not whether you can make a difference to the world and build a wonderful life for yourself while doing so. The question is, do you want to be the type of person who tries? Do you want to be like the smart frog, who relies on the brain that reminds him of his limitations, or the "stupid" frog, whose heart tells him to try anyway?

Maybe you care about food deserts and kids not having access to good food, or maybe it's incarceration of local youth, or maybe, like me, you worry about inaction on climate change. Whatever it is, pick up your placard or call your senator or gather your friends. Don't worry about the second step. Just be too stupid to know that the first step won't work.

Celebrate Your Inconsistency

Did you ever hear anyone say something like "If you care about climate change, why do you heat your house?" Or "If you care about the developing world so much, why don't you give all your money to charity?" As though not completely martyring yourself to your values invalidates the good things you do. As though you have to be either a monk or a scoundrel and there is nothing in between.

The underlying thought is that our inconsistency—or actually, juggling of more than one concern—makes us worse than the completely self-involved people who are, at least, consistent in their self-centeredness. This, of course, makes no sense.

In my hometown, there is a barrier beach that was linked to the mainland by an old rusted bridge over the harbor. Over time, the cost of maintaining the old bridge became so great that people began to think about investing in building a brand-new bridge.

This is similar to what's happening in our culture. The cost of the old consumption-based economy to the quality of human life and the planetary habitat we depend on for our health, happiness, and security is starting to become greater than that of trying to build a new economy that is fairer to people and safer for our world.

But because some people are afraid of change, they say things like "If you want a new economy, why do you participate in the old one? Why don't you give up money? Why don't you give away all your things?"

I think of that old bridge in my town. Eventually, construction on the new bridge began. But as the new bridge was being built, people still drove and walked across the old bridge, of course. No one said, "If you support the new bridge, why do you still use the old bridge?" In fact, the old bridge was hugely important in building the new bridge. Trucks and cranes and materials crossed over the old bridge in order to get to the other side of what would be the new bridge.

That is like our old economic ways of relating to the world. Even if we believe, for example, that a transportation system with far lower reliance on automobiles would be better for people and the world, the systems are not yet in place to allow us to stop using cars altogether. Working for a car-free city but occasionally using a car does not make you a hypocrite. It is like using the old bridge while you are working on building a new one. Inconsistency is just a positive sign that you have begun to change.

Keep Your Eye on Life's Real Prize

One night at a dinner party I attended at a university president's house in Iowa, the guests went around the table taking turns speaking of their life passions and their world concerns. One young college student spoke joyfully about how he had spent the summer delivering secondhand books by bicycle to underprivileged children. He talked about how the job married his passion for economic justice with his love of the secondhand economy and bicycling, which he liked for environmental reasons.

When he finished telling us all about his work, his voice dropped and he said sadly, "I'm really going to miss it." The nineteen-year-old explained that he wanted a family. To support a family, he would have to make a lot of money. To make a lot of money, he would have to go into business. To go into business, he would have to get an MBA. "So I hope, maybe when I'm sixty or so and I retire, I'll be able to volunteer for a nonprofit like that."

The room went quiet with the weight of what the young man said: that he understood and knew what kind of work and life made him happy but was going to sacrifice all that to pursue a standard life approach that would take him somewhere he wasn't even sure he wanted to go.

I told him a story you may have read on the Internet:

On the Mexican coast there lived a fisherman and his family. Every day the fisherman and his children would play on the beach. Then the fisherman would lie on the sand and rest. Eventually, he would go back to his shed, grab his fishing pole, walk down to the water, cast his line, and, in a matter of moments, land a large fish that would feed his whole family for the day.

One day, a business professor happened to be visiting the village, and he watched the fisherman play with his children, cast his line, and, with uncanny skill, land a big fish. Eventually the business professor approached the fisherman. "You should work a little more and play a little less."

"Why?" the fisherman asked.

"Well, if you caught more fish, then you could save up for a boat and some nets."

"Why would I want that?"

"Well, before long you would catch so many fish that you would be able to buy more boats and hire other fishermen to work for you."

"But why would I want that?"

"Well, then you could earn enough cash to build a fish processing plant to supply the entire area."

"Why would I want that?"

"Well, in only about ten years you would have accumulated enough money to stop working."

"Why would I want that?"

"Well, then you could retire and spend your days on the beach playing with your children and fishing."

"Isn't that what I'm already doing?" asked the fisherman.

Be careful to keep your eye on life's real prize.

Trust Your Vision

So you have your idea and you've taken your first steps toward building the new relationships you want to have with the world. You have veered away from one or more of the standard life approaches. Maybe you have even gathered a little bit of energy and success. Great news! This is when the critics and second-guessers arrive. That's a reason for not getting started in the first place, right? Nobody bothers to second-guess you when you're just fantasizing about your great idea.

During my No Impact year, I suddenly found myself invited to go on *Good Morning America* with Diane Sawyer. I was horrified. In hindsight, I'm sure I had an overinflated sense of my own importance, but at the time I was worried I could send people in the wrong direction.

I had no real endorsement other than my own trust in my intentions. I had to go on national television trusting in myself and my vision. Absolutely the hardest thing of all was this: I had to accept that I might be wrong and do it anyway.

Sadly, lots of arguments break out in social change communities about best methods. People tear each other apart as though the scenario is either/or when really it's and/also. We need many shoulders against many doors. What I've learned as I've come to meet so many amazing engaged citizens is that it takes many different strategies and many different styles to make the changes we're hoping for.

So trust your vision. You may find that the biggest risk you can take for the world is to face the possibility of being publicly wrong. And to move forward anyway.

Developing Your Optimism Muscle

My friend Ryan works as a freelancer. He told me, "If I think about how I didn't make enough money last month, I have to get up from my computer and do something else because I feel too anxious. If I think about the ways I did make money last month, I can sit down happily and keep trying." What you think upon grows, they say. You bring toward yourself what you hold in your mind.

It's the same with the world. If I think that climate change is overtaking us and there is nothing I can do, I don't bother speaking, or when I do, my speaking is lackluster. But when I think about the huge number of people I have met who are trying to do something in their own way or the seventy thousand people who have done No Impact weeks, I'm inspired. Being grateful for the good things in the world gives me energy to make more good things.

We can hold a vision of the world we want to live in. We can give thanks for a wonderful world and help that wonderfulness spread to others. This is much better than feeling guilty for what we have and being paralyzed by our shame. This is part of the essence of optimism.

In her book *The How of Happiness,* Sonja Lyubomirsky, positive psychology researcher and author, quotes her graduate school supervisor Lee Ross: "[Optimism] is not about providing a recipe for self-deception. The world can be a horrible, cruel place, and at the same time it can be wonderful and abundant. These are both truths. There is not a halfway point; there is only choosing which truth to put in your personal foreground."

Practicing optimism doesn't mean avoiding the ills in the world or negative information. In fact, research shows that optimists are more, not less, vigilant about risks and threats. Perhaps that is because they can deal with them and don't have to deny them. They are also aware that good things depend on their efforts and they don't have to wait around for the world to fix things for them.

According to Lyubomirsky, if you are optimistic about your goals for yourself and the world, then you are more likely to put effort into achieving them, so optimism is self-fulfilling. You will also keep trying even when you perceive the obstacles and setbacks. Research shows that optimists are more likely to persevere even in the face of difficulty.

In other words, when faced with huge world problems, optimists are both more psychologically resilient and more likely to embark on projects to help the world. It's better for the world and for yourself if you are an optimist. Meanwhile, all it takes to be an optimist is work on maintaining an optimistic outlook—keeping your eye on the clear paths instead of the obstacles.

One exercise scientists have shown to enhance optimism and feelings of well-being is the "best possible selves" diary. I have adapted it for lifequesters and called it the "best possible self and world diary":

Sit quietly with a pen and paper for twenty to thirty minutes. Visualize a future in which everything has turned out the way you wanted both for yourself and for the world, one, five, and ten years from now. You have worked hard and tried your best and the world and your life have changed in the ways you worked for. Now write down what you imagine.

Think positively. For example, visualize a world not *without* fracking but *with* clean energy and drinking water. It's not so much that you want to build a world with no fossil fuel companies so much as you'd like to build a world where communities own their own solar energy plants that offer local jobs and clean air.

It may not come naturally at first, but if you keep this up as a regular practice, studies suggest that you will see changes in yourself and in your life. You will also gain insight into your future and your goals for yourself and for the world. You also, yes, feel happier.

4

THE UNIFYING THEORY OF CHANGING YOUR LIFE AND YOUR WORLD

Here is the unifying theory for life and world change that summarizes everything we have discussed so far and everything we will discuss in the rest of this book. It includes looking for your part in problems and solutions, not getting caught up in false stories, working within your sphere of influence, not wasting your energy on blame, moving toward a happier life and world, not worrying about results, and much more.

> *Give more energy to what is True for you; give less energy to what is not True for you.*

So simple, right? Also, so gentle. Also, sometimes, so hard. In fact, all the scriptures, all the sutras, all the vedas are written just to help us do this one thing: give more energy to what is True; give less energy to what is not True. But we don't need to read all that writing—though of course it can be helpful—because the path can clearly be found within our Selves.

Listen: In 1906, while leading the movement for the rights of ethnic Indians in South Africa, Mahatma Gandhi coined a Sanskrit term for the spiritual force that he believed lay at the root of that movement and, later, of India's independence movement. That term is *satyagraha*. It comes from the words *satya*, meaning "truth," and *agraha*, meaning "insistence" or "holding firmly to."

So satyagraha is what Gandhi called the "soul-force" or the "love-force" that moves toward Truth. Each of us has that soul-force or love-force within us, and the person who follows it—who practices satyagraha—is called a *satyagrahi*. By clinging to Truth, a satyagrahi—what we call in this book a lifequester—automatically moves toward his or her vision for a better life and a better world.

"I discovered in the earliest stages," wrote Gandhi, "that pursuit of truth did not admit of violence being inflicted on one's opponent but that he must be weaned from error by patience and sympathy." Therefore, the pursuit of Truth comes not by trying to change others or the world but by changing the way one reacts to others and the world—by changing oneself.

You don't have to start a revolution. You don't have to change everything at once. You don't have to get down on yourself. You don't have to force yourself. You can't stop violence in the world by doing violence against yourself. Instead, all you have to do is begin to trust your Truth. Give energy to what is True (or loving) and starve energy from what is not True (or not loving).

The purpose of the rest of this book is to help you decide what is True and loving for you.

A Free Workbook and Extra Resources

In order to help you work through this book, I have created a workbook that includes worksheets, resources and additional exercises. These include lists of webpages that will help you as well as ways to answer many of the questions that will come up for you as you look for a life that is True for you. You can download the workbook for free by going to **www.colinbeavan.com/HowToBeAliveResources.**

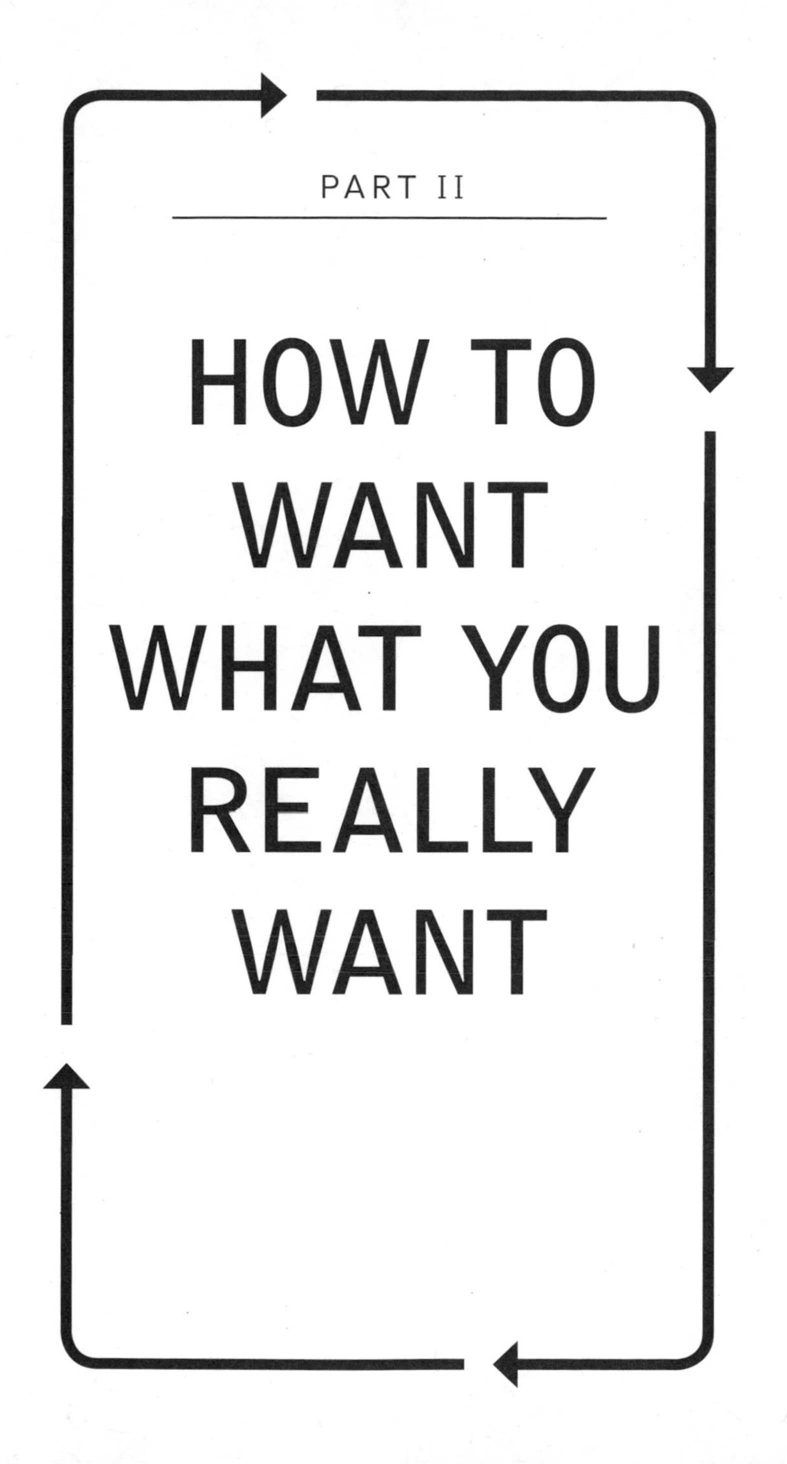

PART II

HOW TO WANT WHAT YOU REALLY WANT

5

THE COOLEST PART OF BEING A LIFEQUESTER

If you came to this book because you are worried about the world and wondering how to help, you may wonder how "wanting what you really want" has anything to do with it. Isn't that exactly what we are trying to get away from? Everybody just wanting what they want and not thinking about others or the world at large? Well, no, because many of us aren't actually focused on what we really want but on what we think we want. If we could get past the false ideas and stories about what we want, things might work a whole lot better.

I'm going to explain that later. "What you really want" is something we will explore in depth. But for now, let's assume it isn't more stuff and money but what might *really* make you happy—being True to your Self. That sounds all woo-woo, but I promise I'm going to marshal the science.

For now, the coolest thing to say about being a lifequester—a satyagrahi or truth-follower, as Gandhi called it—is the principle that *if you pursue Truth for yourself, you cannot help but pursue Truth for everyone.* That means that if you really cling to the truly right way for you—if you want what you really want—it is going to be the right way for you to help others, too. Saving yourself saves the world; saving the world saves yourself.

This can be true if you transform your whole life in one big dramatic step or if you change any one of your relationships with the world.

A historical example of how pursuing Truth for yourself leads to Truth for all:

When Gandhi led the independence struggle for India, he repeatedly made clear that the oppression of the Indians by the British *undermined the human dignity of the British soldiers and administrators at the same time that it undermined the dignity of the Indians.* The nonviolent liberation of Indians from British rule would also liberate the British from the waste of life energy (time and effort) and the psychic costs of perpetuating oppression and violence—from lives based on non-Truth.

Following Truth for the Indians automatically meant following Truth for the British. Following Truth for oneself breeds Truth for others.

Pursuing the Good Life—a life based on Truth or wanting what you *really* want—for yourself helps to liberate others to find Good Lives for themselves, too. This means you can move forward not in a spirit of antagonism, competition, and anger but in one of love and compassion. To pursue the Good Life is automatically to be a leader. It is to trust that following your own True Path helps others. Because Truth is Truth. That's another cool idea Gandhi had.

You might refuse to buy meat produced by the intensely cruel American factory farming system because you don't care to ingest or have your children ingest the stress hormones and other chemicals in conventionally produced meat. In refusing it, you also stop giving energy to the indignity wrought on impoverished farmworkers when they are forced to put aside their innate compassion toward other sentient beings in order to make a living. Truth breeding Truth.

This is how Martin Luther King Jr. put it in a 1965 speech at

Oberlin College: "All mankind is tied together; all life is interrelated . . . Whatever affects one directly, affects all indirectly. For some strange reason, I can never be what I ought to be until you are what you ought to be. And you can never be what you ought to be until I am what I ought to be—this is the interrelated structure of reality."

A fundamental principle of the Good Life for a lifequester is: When I become more, you become more. When you become more, I become more. When you become what you are really meant to be, you cannot help but benefit the world.

How One Person Becoming More Helps Others to Become More: The Story of Rabbi Steven Greenberg

Suppose, first, you are a man and you discover you have a passion for religious study and you decide to become an Orthodox rabbi. Suppose, second, you realize, after years of struggle, that there is nothing you can do about the fact that you are attracted only to other men.

Doesn't that make you the worst kind of contradiction? Don't religious traditionalists hate gays? Don't gays hate religious traditionalists? Where on earth is a gay Orthodox rabbi supposed to fit?

The thing is, if you can figure out your path through your supposed contradictions, if you have the courage to accept all of who you are and to follow your individual Truth, then maybe there is a chance you can help other people who share your struggles. This is what we mean when we say one person becoming more makes other people become more.

My lovely friend, spiritual adviser, and rabbi (though I am not Jewish) Steven Greenberg is the author of *Wrestling with God and Men: Homosexuality in the Jewish Tradition* and appears in the documentary *Trembling Before G-d.* Through his organization Eshel, he has helped thousands of lesbian, gay, bisexual, transgender, and queer Orthodox youth and adults accept their Truth, reconcile with their communities, and save themselves from rejection, depression, and even suicide.

Here is what Steven told me about his story:

I grew up in a Conservative Jewish family. When I was fifteen, I had begun to regularly walk the mile trek to the synagogue to avoid being driven by automobile, something traditionally prohibited on the Sabbath. One Saturday morning, it was drizzling and my mother insisted that I get in the car. As an act of teenage defiance, I refused and walked four blocks in the rain to the Orthodox synagogue instead. The rabbi there befriended me, invited me to Sabbath lunch, and proposed that I learn with him every Shabbat afternoon and, surprisingly, I did.

I loved learning Torah. The sprawling rabbinic arguments and the demand for meaning everywhere thrilled me. Membership in this small devoutly religious community connected me in new ways to my Jewish heritage. Also, I didn't realize it then, but negiah, *the prohibition to embrace, kiss, or even touch girls until marriage under traditional religious law, was a perfect mask, not only to the world but to myself.*

In 1976, at age twenty, I began my rabbinical studies in Jerusalem. I would wake up in the morning to the sound of a particular fellow male student showering next door. I was disturbed to discover that it made me feel excited and that I wished I could see him naked. I suddenly realized I had been feeling this way toward men ever since I could remember.

I was scared. I didn't even know what "gay" was. I had heard the word faggot *but I thought it meant "bad at sports." I rode on a bus for an hour and a half to a remote corner of*

Jerusalem to consult with the great rabbinical sage Rabbi Yosef Shalom Elyashiv. I sat in an anteroom with others who waited for hours to speak with him for a few minutes.

My turn came and I took my seat across from him. I was so nervous. I said in Hebrew, "Rabbi, I am attracted to men and women both. What should I do?"

"You have twice the power of love," he said. "Use it carefully."

I really expected more. I said, "Master, is there anything else?"

He smiled. "There is nothing more to say."

I was so relieved. I knew he wasn't saying it was okay for me to have a boyfriend and I wasn't looking for that. I wasn't ready anyway. I just wanted to know that my inner life didn't make me a monster, and the rabbi had assured me it didn't. I was okay.

Still, for fifteen years after that I tried to create a life that would allow me to eventually marry a woman. Meanwhile, I secretly saw men. I forced the Truth of myself into the darkness, and in that darkness, I now realize, I did harm to myself and others—particularly the women who couldn't understand why dating me didn't work out. I didn't know it then, but to be True to others I would have to be True to myself.

During the prayer service on Yom Kippur, we read the chapter in Leviticus on sexual violations including the famous verse: "With a male you shall not lie, in the manner of lying with a woman, it is an abomination." Every year, I would hear this reading and cry. By Yom Kippur 1995, I had no more tears left. I realized that I had to come to understand the text in a way that did not make me hate myself.

Leviticus can be read as hateful of homosexuality only if it is taken entirely out of context of the rest of the Torah, which teaches us, among so many other wonderful things, to love—not to judge—our neighbors. I began to wonder if perhaps the rabbis misinterpreted the Leviticus verse because people like me did not tell the truth of who we are. The rabbis had never really heard our stories. Without hearing from us, how could they interpret the text in a way that accommodates the entirety of human experience?

I got a two-year stipend to study rabbinic attitudes toward sexuality in Israel, and that is when the seeds of my book were planted. Meanwhile, I helped a group found the Jerusalem Open House, the gay and lesbian community center of Jerusalem. Toward the end of my stay, they asked me to publicly come out in a newspaper interview to help promote the center's opening. I said no. But then I ended up sitting on the airplane next to the journalist who was supposed to interview me and I decided it was fate.

A week later, in March 1999, I was suddenly proclaimed to be "the world's first openly gay Orthodox rabbi." Then it snowballed. A filmmaker named Sandi DuBowski interviewed me for his film Trembling Before G-d, *about gay and lesbian Orthodox Jews trying to reconcile with their faith. The movie came out in 2001, and in 2004, after nearly ten years of study, thinking and writing, my book came out.*

Almost overnight, I seemed to become one of the world's spokespeople for gay and queer Orthodox Jews. Since I was a rabbi, it suddenly became my job to talk to other rabbis and communities about how we can continue to love and support our own brothers and sisters and mothers and fathers and sons and daughters even after they are brave enough to tell the truth about how they love.

I filled this role before I had even had my first openly gay relationship. It was only when I so publicly came out, in 1999, that I finally met my life partner. Sandi brought a friend to Shabbat lunch at my house. The guest was Steven Goldstein, the opera singer and professor who later became my husband.

In 2010, our daughter Amalia was born—her name means "God's work." By then, I had worked through my fears and self-rejection and was ready to be a parent. As important, I had trust in the vitality of my religious tradition to move forward.

In 2011, I founded Eshel. Our mission is to support and care for LGBT Orthodox Jews who are struggling to be themselves without walking away from their communities. We help them and the people who love them to tell their stories in their communities so that their communities can learn to embrace all people.

Recently, for example, a teen wanted to change biological gender. The teen proclaimed to the parents, "I'm having a bat mitzvah, not a bar mitzvah." Since then, their rabbi won't let this child into the synagogue. So through Eshel, I gathered twenty rabbis on the phone with the mother of this teenager to hear the family's story. It is my privilege to work with many families like this.

It is a crazy life I have, in some ways, but such a wonderful one. I mean, a gay Orthodox rabbi! It's almost comedic! But following the Truth has brought me a beautiful family and community of friends. There is a lot of love. I get to study and write and think about what concerns me in the world. I get to watch people being incredibly brave on their own journeys and to help facilitate that. And I get to feel that the suffering in my own life has meaning and purpose. This is a great satisfaction to me.

6

WHAT ARE PEOPLE FOR?

When we talk about wanting what we really want, what do we mean? How *exactly* do we want to be in the many relationships with the world that make up our lives? What is it that we *really* want from those relationships, both for ourselves and for others? What is it that we should be striving to accomplish in our short time on planet Earth?

In other words, what is it exactly that the lifequester is questing for? What is it exactly that the Truth-follower is following? What is the "ought" that Martin Luther King Jr. referred to when he spoke of becoming what we "ought to be"?

These are questions we will have to embrace if we hope to find better lives for ourselves and our communities.

What are people for?

What a Wise Man Worries He Hasn't Accomplished at the End of His Life

In the late 1700s, every day rabbinic students crowded to the *beth midrash*—the house of Jewish religious study—in Anipoli, Ukraine, to learn from the mystical Hasidic rabbi Zusya.

One day, in 1800, Zusya did not appear at the usual hour. The students waited and, finally, rushed to Zusya's house to check on him.

They found the rabbi in his bed, too ill to get up. He was dying and he was terribly upset.

His students felt confused. "Haven't you taught us that all living things must die, that it is natural? Why are you upset?"

"Yes, it is natural to die. All living things must die," he said.

A young student tried to comfort him: "Then you have no need to be upset. You have lived a life with as much faith as Abraham. You have followed the commandments as carefully as Moses."

The rabbi summoned his strength to answer his students.

"Thank you," he said. "But that is not why I am upset. If God asks me why I didn't act more like Abraham, I'll say because I am not Abraham. If he asks me why I didn't act more like Moses, I'll say I am not Moses."

Then he paused and looked at his students. His eyes filled with tears.

"I am upset because I have been wondering, if God asks me why I didn't act more like Zusya, what then will I say?"

How Another Wise Man Figured Out What Direction to Follow in Life

During the life of the Buddha, the thousands of monks and students who followed him—the *sangha*—would gather regularly at various holy places and monasteries for meditation retreats and trainings. At these gatherings, whenever Buddha was about to give a talk, a flag would be raised outside the gate of the dharma center to alert followers that it was time to gather.

After Buddha's death, a large group of his "enlightened" students convened to formally record these teachings into what are now called the *sutras*—the Buddhist gospels. The head of this group was Mahakashyapa. Buddha had given his bowls and gold-embroidered robe to Mahakashyapa, more or less crowning him as Buddha's dharma heir (the most advanced and senior follower).

Among the other senior students who made their way to this historic assembly was Ananda, even though he had not achieved "enlightenment." As Buddha's personal attendant, however, Ananda had been present for more of Buddha's talks than any other follower. He also had a reputation for having a near-miraculous memory, and people believed he could recall every word Buddha had ever said.

Ananda thought that since he knew the most about what the Buddha had said, he would probably be the most important contributor at the convention—enlightened or not. Yet when he arrived, Mahakashyapa blocked Ananda's entry. Ananda was furious. What right did Mahakashyapa have to stop him?

"Besides for that stiff old robe, what did you get from the Buddha, anyway?" he asked.

He was too diffident to say more, but what he might have added was, *You may have that golden robe, but what directions do you think the Buddha gave you that I never heard? What makes you think you know more about Buddha's path than I do?*

Maybe Ananda thought Mahakashyapa looked down on him for being Buddha's servant, but that wasn't it at all. What Ananda apparently didn't yet realize was that the root of Buddha's teaching was not to follow some dead old directions that come from outside yourself, not to follow some other person's path, but to follow the thing that all of us have within us—the True Self, the You within you, or what the Buddhists might call the Buddha within you.

"Ananda!" Mahakashyapa shouted.

"Yes, sir!" replied Ananda, perhaps thinking he was about to get rebuked for his rudeness.

"Knock down the flagpole in front of the gate!"

That wasn't what Ananda expected to hear at all. *Knock down the flagpole in front of the gate?* The flag was the sign that Buddha was about to give a talk. By knocking down the flagpole at the front gate, Mahakashyapa seemed to be saying, *Never mind all those dumb words Buddha said. Knowing Buddha's words doesn't bring you enlightenment.*

This really confused Ananda. What Mahakashyapa had said was like saying *Who cares what Jesus said?* or *Who cares what Moses said?* or *Who cares what Muhammad said?*

Besides, Ananda had always been praised for knowing the Buddha's words. Why did his word-knowledge suddenly not matter? If words were unimportant, why had everyone listened to Buddha for so many years? And why was everyone gathering to record the words he had said? What did Mahakashyapa have that Ananda did not? What was the true teaching?

So Ananda went to the mountains to sit in meditation. For seven days and nights his confusion gripped him. *If the words and directions don't count, what does? What is the Truth?* He did not sleep or eat. He just meditated. Suddenly, something shifted within him. He got up from his seat and marched back to the assembly. When he arrived, Mahakashyapa again blocked Ananda's entry. He put him to another test.

"If you can come in without opening the door, you may come in," Mahakashyapa said, pointing to the closed door behind him. "If not, you may not come in."

Ananda pushed past Mahakashyapa, opened the door, and went in. *Never mind your mumbo jumbo,* he might have said, *you guys need my help!* Ananda finally trusted himself. Ananda had learned to act like Ananda.

Mahakashyapa followed Ananda through the door, smiled at the assembly, and said, "Ananda has arrived! We may now finally begin recording the sutras!"

Why It's So Great That We Aren't Dead Yet

There is an old *New Yorker* cartoon I have often mentioned when speaking to audiences across the United States and Europe. In it, an old man is lying on his deathbed, gasping his last breaths. His wife is holding his hand, saying her final good-bye. There is a look of regret on the dying man's face and his wife is leaning over him to hear his precious last words.

Here is what he whispers: "I only wish I had bought more crap."

Audiences laugh and laugh when I tell them about this cartoon. Most of us believe the quiet voice inside us that tells us that what is most important to us is not the "crap" we buy—not stuff. What makes the joke funny, though, is not so much that the old man is wrong but that we identify with his wrongness. We worry that we might be a little too much like him, that our lives might be flying off in the wrong direction.

So many of us have organized our lives so that most of our time is spent earning money for stuff, buying stuff, using stuff, and disposing of stuff. We know that spending our lives chasing after stuff isn't really who we are or what will make us happy or how we can help the world. We have even seen the scientific research that says so. Yet somehow, we are caught in what scientists call the *hedonic treadmill*—the tendency to keep chasing after experiences that offer only a brief spike of happiness so that we have to keep trying over and over again and never really get anywhere.

Nor can any of us expect to find a path to individual and world happiness simply by jettisoning the standard life approaches and then blindly following a new, better, different set of directions. You can't just cross out "work hard and spend money" and pencil in "emphasize relationships and do service." Because what about joking around and having fun and going on adventures and being kind of naughty? What about having a little edge (or a lot of edge)?

It's not that the directions about relationship and service aren't better than the directions to work and spend, because they are. But they are still, after all, directions. Instead of trying to connect a better set of somebody else's dots, don't we want to wake up and connect our own dots? Don't we want to wake up to that place inside us from which our own direction might come? Because someone else's better directions may lead us down a better path, but following directions, we might still only sleepwalk along it.

Like Ananda, we were not born to follow someone else's path. Like Zusya, we won't regret that we weren't more like someone else but that we weren't more like our Selves. We know this, and this is why the *New Yorker* cartoon is funny: because we know the dying man is saying exactly the opposite of what we will be saying on our deathbeds. We will not be like him. We may be worrying, but it won't be about the stuff.

And so, when I've stood in front of these audiences, in front of these many thousands of Americans and Europeans, because they worry what they will say in that hospital bed, for all the laughter, there is an undertone of sadness and regret. "You feel a little worried," I say to them. "That is so good! That worry is the feeling that tells you it is time for a change."

The bad news, and you probably know this, is that to find a way to your Self you will have to do some work and accept some confusion. The good news is that right here and now, as you read this book, as I write these words, you are not yet dead. You still have time.

Want What You Really Want Tip #1: Clean Out Your Mind Clutter—Take a Media Fast

Most of us have spent many years not listening to our Selves, and thus it is hard to tell what there is to hear. When you add the external chatter of contemporary consumer-driven life, it's even harder. Some studies show that we are exposed to as many as ten thousand advertisements a day. How are you supposed to be able to hear that some quiet inner voice is telling you what you *really* want when ads on TV, on the Internet, in movie theaters, on magazine pages, and even on your cell phone are telling you what you *should* want?

Back in 1929, Charles Kettering, the chief engineer for General Motors, was in charge of making people want to buy new cars even when their old cars still worked perfectly. In part, his job was to make people dislike their old cars. He wrote: "If everyone were satisfied, no one would buy the new thing because no one would want it. . . . The key to economic prosperity is the organized creation of dissatisfaction."

That's why media is deliberately designed to be a dissatisfaction creation machine. Media shows us rich people with better-looking lovers and cooler stuff and bigger houses and more exciting lives who never seem to have to work. Media creates an unrealistic view of life that we then compare ourselves to. It makes us feel frightened and insecure—and like we need more stuff.

The noise of the dissatisfaction machine is so loud that we can't hear the Self who tells us otherwise.

One way to begin to learn what you really want is to turn the dissatisfaction machine off. Turn the noise off. Take a media retreat or a media fast. Research shows that people on media fasts enjoy fewer cravings, better sleep, more activity, more free time, less stress, and other benefits.

Here's how to do it:

Adopt an attitude of experimentation—this isn't about depriving yourself but about seeing whether a new way of being might actually make you feel better.

Ask a friend or small group to join you in your fast and check in with them once a day about how it's going.

Choose a period of time that you can succeed at—a week, two weeks, a month?

Decide which screens you will turn totally off—for example, you may need to use your phone for work, but no one really needs to watch TV. What other screens can you turn completely off?

Set limits for the devices you can't totally turn off—for example, check e-mail only at certain preset times, or make all web pages you don't really need to look at off-limits.

Choose things you really want to do with the newly available time—cook a meal for all your friends, or dust off your guitar and learn a new song.

Keep a brief journal of your state of mind and what you discover.

When your fast is over, make a conscious choice about how much time you want to spend on screens moving forward.

Choose only media that reinforce the values you want to emphasize.

Go on another fast when the screen time creeps out of control again.

7

HOW TO KNOW HOW MANY REGRETS YOU WILL HAVE ON YOUR DEATHBED

Do this. Just for fun. It's going to take less than ten minutes. Grab a pen and paper. Draw a big clock face with the numbers one to twelve.

Now imagine you have twelve completely free hours to build your life exactly as you want. You get to split the twelve-hour day into parts and decide how much time you want to spend on each of the eleven types of aspirations listed opposite. You can devote as much or as little of your twelve hours as you want to each area of life. In fact, you can choose to completely ignore one or more domains.

One bit of good news in the fantasy world of this exercise: You don't have to know what you will do to achieve your ambitions. You just have to decide how much of your twelve hours you want to spend on each one.

This list of universal human goals* and others very similar to it come from a large body of cross-cultural research showing that people's aspirations turn out to be very similar across nations, cul-

* List adapted from Frederick Grouzet et al., "The Structure of Goal Contents Across 15 Cultures," *Journal of Personality and Social Psychology* 89, no. 5 (2005): 800–816.

tures and economic circumstances. What varies is not the goals people care about but the relative priority people assign to the various goals. By the way, don't be distracted by the *I* or *E* after each goal for now. I will explain that later.

Affiliation (I): to have satisfying relationships with family and friends

Community feeling (I): to improve the world through activism or generosity

Conformity (E): to fit in with other people

Financial success (E): to be wealthy and materially successful

Hedonism: to experience much sensual pleasure

Image (E): to look attractive in terms of body and clothing

Physical health (I): to feel healthy and free of illness

Popularity (E): to be famous, well-known, and admired

Safety (I): to ensure bodily integrity and safety

Self-acceptance (I): to feel competent and authentic

Spirituality: to search for spiritual or religious understanding

Divide the clock face according to your priorities. Order doesn't matter. For example, if you want to spend an hour and a half on your first activity, draw a line from the center of the clock to the 12 and another one from the center to halfway past the 1. Write the name of the aspiration you are dedicating the time to in the resulting wedge. Continue on until you have used your twelve hours. Make any changes you want.

When you are done, use lines to shade in all the areas associated with aspirations that have an *E* after them, as in the example shown on the following page. Use crisscrossing lines to crosshatch the areas associated with aspirations that have an *I* after them (as I said, I will explain the meaning of the *E* and *I* later). Don't shade the areas associated with neither *E* nor *I*. You should be left with a kind of pie chart representing the different kinds of things you want in your life.

With this exercise, you have already accomplished something valuable. You have effectively shown yourself what your goals in life are. That's important. The next question is whether your goals are in line with your True Goals: whether they will help you become what you "ought to be," whether they will make you happy if you achieve them, whether they will help you help the world, and what their order of priority in your life does to your happiness in the first place.

In other words, do you want what you really want?

My friend Tim Kasser, a professor and author of the book *The High Price of Materialism,* helped me design the clock test you just took.

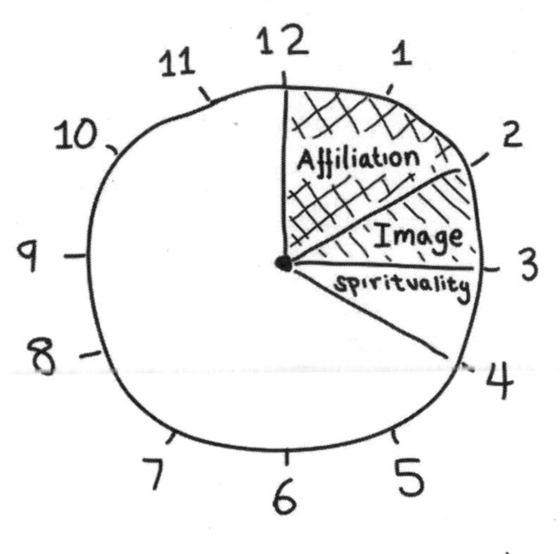

Divide a 12-hour day according to how much time you want to spend on each goal.

Tim is perhaps the world's top expert on the relationship between materialism and human (un)happiness. One major area of his research is the question of whether the goals you prioritize in your life make you happy.

As early as 1993, as a doctoral student with his supervisor Richard Ryan at the University of Rochester, Tim began using a tool he called the Aspiration Index to scientifically determine how important an individual's materialism and financial success (*extrinsic* goals, labeled *E* in the list on page 105) were in comparison to other values like personal growth and acceptance, family life and friendships, and the desire to make the world a better place (*intrinsic* goals, labeled *I* in the list).

Back then, Tim administered the Aspiration Index to 316 University of Rochester undergrads. The students also filled out four other questionnaires to determine their feelings of well-being and distress, their vitality, and their levels of depression and anxiety. Using statistical analysis, Tim determined that the students who held financial success as a relatively central goal were less happy and less psychologically healthy.

This result has been confirmed over and over again in countless tests using the Aspiration Index in numerous countries over the last twenty years. Research shows that subjects who prioritize materialistic goals are more likely to suffer from substance abuse, carry weapons, be involved in vandalism, report anxiety and depression, watch more TV, have fewer friends, have shallower relationships, suffer from more psychiatric conditions, be less happy, and so on.

In other words, regardless of your financial status or background, the more materialistic your goals are, the less happy you are (though there are good reasons why some of us are materialistic—more on that later). Not only does making money and owning a bunch of stuff not necessarily make you happier, but orienting your life around making money and owning a bunch of stuff correlates with being actively unhappier.

Simply put, scientific research shows that wanting money and stuff makes you unhappier than wanting to grow and contribute.

Why is this so? Why would chasing after money and stuff make you unhappy?

Because chasing after these materialistic goals takes time and effort away from the nonmaterialistic goals—like relationships, service to community, health, etc.—that actually *can* make you happy. Materialistic people choose low-satisfaction jobs and relationships with people who make them look good. Their goals guide them to choices that make them less happy.

Chasing after stuff is a little like being a stray countryside dog who chases after cars. He exhausts himself with the running, bites into tires he can't eat, and puts himself in harm's way. What good does it do him? None. He just chases after the next car. Then the next. And all the time the stray dog is chasing cars, he is not doing the productive things that might actually help his life—like maybe catching dinner.

If you waste your life chasing after the stuff that doesn't make you happy, then you will end up short of the things that do. In other words, you might say that, since it can make you so unhappy, *wanting money, stuff, and status is the opposite of wanting what you really want.*

Here is what Tim told me about the clock test you just took:

"Most people would not say that money is the most important thing in their lives. So the problems people have in their lives don't come when money and stuff is their first or second priority. The problems for people come when money and stuff is in the middle or above in the eleven items in their priority list."

We will discuss how to potentially change your personal goals later on. First, we are going to look at the science behind why materialistic pursuits don't make us as happy and what actually does.

Meanwhile, if the area shaded with lines on your clock covers anywhere near the same amount of space as the area that is crosshatched—if your materialistic goals are of equal or greater importance than your intrinsic goals—it may well be time to take action to avoid the fate of the dying man in the *New Yorker* cartoon.

Want What You Really Want Tip #2: Look More Deeply—What Would the Money, Looks, and Fame Bring You?

Sometimes we run so fast on the work-to-spend treadmill that we never stop to ask what we want the money, stuff, and fame for. How do we imagine they will make us happy if we get them?

Tim Kasser writes: "The truth is that money is good only for buying food, shelter, safety, and other necessities; it can never really buy self-esteem, love, or freedom." Similarly, you can work your rear end off for good looks, and it may attract more people around you, but what kind of people will you attract who wouldn't have liked you anyway?

Write down the needs you imagine will get filled by fame, money, or looks. What will getting those things make you feel, do you imagine?

Remember times when you achieved a materialistic goal. Did the new car, the new raise, the new gadget actually satisfy any of the needs on your list? How long did it make you feel good?

Ask yourself, is it possible that the materialistic goal is actually the long way around to getting what you really want?

If your list of needs includes things like friends, less anxiety, lovers, laughter, fun, or adventure, can you get them more directly?

Write down some strategies for finding those things directly so you don't have to jump through so many materialistic hoops. For ways to create your own personal community, for example, see Part IV: Finding Your People.

8

WHAT SCIENCE SAYS ABOUT WHERE YOU WILL ACTUALLY FIND PURPOSE AND MEANING

A Quick Review of What We've Learned About Wanting What You *Really* Want

1. The coolest part of being a lifequester is that if you chase after what you *really* want—follow what is True for you—then, according to Gandhi and Martin Luther King Jr. (and the science we are about to discuss), you cannot help but somehow help others.

2. According to Zusya and Ananda, what you really want—your Truth—cannot be found in a set of directions or by following someone else's path. You have to look inward.

3. Research shows that chasing after fame and fortune, even if you do it in your own individual way, actually makes you actively unhappier and your life worse. So wanting money and stuff is not wanting what you really want.

Now it's time to find out a little more about what *does* count as wanting what you really want.

A Monkey Who Did Things Just for the Hell of It

Here is a true story about a rhesus monkey who happened to live—poor thing—in a primate laboratory at the University of Wisconsin–Madison back in the 1940s. As far as I know she didn't have a name, but let's give her one: Jane. Along with a psychology professor named Harry Harlow, Jane and seven other rhesus monkeys helped establish the science behind human motivation, purpose, and meaning.

One day, in 1949, Harlow put a mechanical puzzle in Jane's cage and also in the cages of the seven other rhesus monkeys. The puzzle consisted of hooks and clasps and hinges. With the right series of moves you could unlock the puzzle and open it or close it and lock it up again. Harlow put the puzzles in the monkeys' cages as preparation for a coming experiment in which he would reward them with raisins for learning to operate the puzzles.

Until the raisins came, went the theory of the time, Jane would have no reason to play with the puzzle. Back then, psychologists subscribed to a model of behavior known as *behaviorism*. The model said that all behavior was driven by one of two things: (1) the internal biological needs for food, sex, sleep, or water; or (2) the external rewards and punishments—pleasure and pain—that come from our environment. With no raisins to induce Jane's interest, behaviorist theory predicted that she would investigate the puzzle just long enough to learn it offered neither pleasure nor pain, and then ignore it.

That isn't what happened.

Instead, Jane totally obsessed over the puzzle. She spent days fiddling with it and figuring out how to take it apart and put it back together. She loved it. So did the other monkeys. But why? Science had no explanation for monkeys doing things just, more or less, for the hell of it.

Watching the monkeys, Harlow reasoned that since Jane and her pals worked on the puzzles without any promise of an *external* reward, they must have found working on the puzzles *internally* rewarding. Jane seemed to have some *intrinsic* psychological need to attempt to solve the puzzle. Satiating that need felt pleasant, just like satiating the need for food or water. But what was the need?

Harlow postulated the presence of an as-yet-unknown motivational drive that made the performance of certain tasks pleasurable. In other words, primates do some things just because it feels good to do them—just for the hell of it or because they are, in a certain way, fun. Performing some tasks and rising to certain challenges are "intrinsically rewarding."

Each of us has things that we do because they happen to appeal to us. Because they are kind of fun. We feel good when we do them. We are motivated to do them without the presence of an "extrinsic reward." Just because that's who we are.

But here comes the second big surprise. Harlow figured that if the monkeys loved doing the puzzles just for fun, they would love doing them even more if they also got a form of reward—the previously mentioned raisin—when they did them right.

Now, Jane and the other monkeys definitely loved getting raisins, yes, but the unexpected result was that when the monkeys started doing the puzzles for the sake of getting the raisins, they all performed worse than they had before. They performed with less, rather than more, enthusiasm. You could say that once the monkeys got hung up on trying to get the raisins and the anxiety of getting them or not, doing the puzzle itself stopped being fun.

Harlow discovered that the monkeys' desire for the raisins was less of a motivation than the desire to do the puzzle for the fun of it. At the same time, the desire for the raisins disconnected the monkeys from their internal spark, from the innate curiosity in the puzzle that made it fun and initially made them want to solve it.

In other words, Jane couldn't help but orient herself around extrinsic rewards—life's shiny baubles—but chasing after them just wasn't as satisfying as doing things just for the hell of it, for fun. So with the promise of a raisin, she worked on the puzzle but not with the same gusto as when it was her internal desire to do so. When she started doing stuff for external rewards instead of because she was internally passionate, she kind of lost her edge.

Sound familiar?

It's like all your life you dig chasing rainbows, but then someone tells you there is treasure at the end of them. Suddenly, you forget how much you dig chasing rainbows and instead you start looking for them to get the treasure. Then you get to the end of a rainbow and discover the real treasure is not there, so why bother chasing rainbows? Somehow, the passion you used to feel for rainbows just isn't there anymore. This is the problem with living for the treasure. Life is less fun.

Harlow's discovery of intrinsic motivation—the desire to do things because they happen to appeal to us—got almost no attention when he published it. Scientific orthodoxy at that time believed the behavior of animals and people could be entirely predicted based on rewards and punishments. It was as though science wanted to be able to reduce people to a bunch of easily pushed buttons so they could be slotted into the automated factories of the burgeoning new industrial complex.

Harlow's results didn't fit.

That doesn't change the fact that many of us can't be happily slotted into the standard life approaches of modern society. Most of us would rather do stuff because it feels authentic to us as individuals than because of the reward of some carrot or the punishment of some stick. That's the conclusion Harlow's work with Jane points toward.

Want What You Really Want Tip #3: Figure Out Who You Are by Doing Things for the Fun of It

Ever notice how much more at ease you are and intelligent you seem when talking to a prospective date or interviewing for a prospective job you don't care that much about? Your jokes are better. Your conversation is smarter. But then when you try to talk to someone you really are attracted to, you trip over your words and can't think of anything to say?

That's because you're trying to get the metaphorical raisin—the extrinsic reward. You want the date or the job interviewer to like you and you are trying to figure out how to make that happen instead of just having fun, amusing yourself, and doing your own thing. Attachment to results divorces you from your spark.

It's the same with playing pool or basketball or making art or playing music or living life. If we feel pressure from ourselves or someone else to perform in order to achieve some reward or result, we get nervous, perform worse, and have less fun.

There is a reason people tell us to "just be ourselves." It is because when we are trying to figure out how to avoid the stick or get the carrot, we kind of split ourselves from that part that just does things for the joy of it. And that's the part that makes our lives work well and feel worthwhile.

But some of us have been chasing after extrinsic rewards for so long that we no longer even know who we would be or how we would act if we weren't.

To reacquaint yourself with wanting what you really want, take an hour, a day, or even a week to do exactly what you feel like doing, just for the hell of it. What you do just for the hell of it—just for fun—will tell you a lot about who you really are. You'll probably find you are kinder to people than when you are chasing a carrot, too.

Here are some hints:

1. Schedule an hour or two to just be. Choose a meaningful amount of time but not one that will make you feel so stressed that you

can't indulge in it. During this time you will not be allowed to do anything to achieve a result (also, no screens!). This will be a period of pure play, just doing whatever feels satisfying in the moment.

2. Prepare for this time by making a list of the worries and concerns that have stopped you from taking this time before. Think about exactly how you would feel if all those things were taken care of. Write down the pleasant feelings and physical sensations that would come with that.

3. When the scheduled time arrives, sit down and imagine all those worries and concerns are taken care of. Try to imagine yourself feeling all the associated pleasant feelings and sensations. In this way, you trick your body and mind into allowing yourself to play. Repeat this whenever you feel anxiety creep in.

4. Do whatever you feel like! Let yourself enjoy. Don't worry if it feels awkward the first time.

5. When the period is over, write down the things you did that you really enjoyed.

6. Repeat this exercise a few times over the course of a few weeks.

7. At the end, you should find that you have a list of things you would do just for the hell of it, things that are internally rewarding to you.

8. Congratulations! You have allowed yourself to want some of the things you really want.

Why We Ask, What Are People For?

It is not an esoteric question.

It is hugely practical.

In fact, the most important practical question there is, for a number of reasons.

First, if we are going to organize our relationships with the world to give us security and meaning, we have to know what it is that actually gives us that security and meaning—what we are for; or to put it another way, what we really want; or to put it still another way, what makes us truly happy.

Second, how are we supposed to feel purposeful without having some sense of what we are for? (Actually, it is possible to trust that you and your life are purposeful without explicitly knowing your purpose, but you get my point.)

Third, if we hope to motivate ourselves and others to help the world, it could help to figure out ways of helping the world that make us happy, too. Then we'd really want to do it, right?

Fourth, shouldn't all governments, corporations, civic structures, economies, and societies aim to help the least and the most of us become what we "ought to be"? If so, shouldn't we know what it is we ought to be—what people are for—in order to wield our influence over those institutions correctly?

THEORY OF HUMAN PURPOSE #1: WHAT AN ANIMAL TRAINER WOULD SAY

One view of human nature in our culture comes out of behaviorism, the school of psychological thought prevalent at the time of Harlow's experiment with Jane. Many behaviorists—like the late, famous B. F. Skinner—believe that human motivation consists entirely of chasing after experiences that feel good and running away from experiences that feel bad.

This view suggests that we are just flesh-based robots that move away from what we don't like—our aversions—and toward what we do like—our desires. Our brains act as complicated computers that try to predict what will feel good and what will feel bad in a variety of future circumstances.

A kid learns things only because she is motivated by the prospect of good grades. A worker works only because he is paid. A person does right by the world only because she is rewarded. According to the behaviorists, the trick to motivating a behavior is to figure out how to make people feel pleasure when they do what you want them to do and discomfort when they don't.

Since the human race is doomed to continue pushing whatever

lever makes it feel good, even if we end up destroying each other and our planetary home, the only way to make sure we do the right thing is to make it so that pushing the lever that saves the world also makes us feel the best.

Society needs to be organized into a sort of ongoing animal training school that automatically rewards people when they help the group and punishes them when they harm it. This is the only way to save the world, because there is no doing things "just because." There is no free will. There is no "human purpose."

THEORY OF HUMAN PURPOSE #2: WHAT A MIDCENTURY VIENNESE NEUROLOGIST WOULD SAY

Another of our culture's prevalent views of human nature comes from the theories of Sigmund Freud. He argued that we are not quite automatons, as the behaviorists suggested, but are instead in a constant struggle between our basic animalistic drives and the need to subdue them in order to live in society. We don't necessarily like society or desire to serve it, but we understand its advantages to our personal survival.

According to this view, we can, with great exertion of our will, conquer or "sublimate" our desires and aversions in order to make ourselves acceptable to others. We would all really prefer to be acting like Mr. Hyde, but as long as anyone is watching, we go around acting like Dr. Jekyll to make ourselves acceptable. This view tells us that if no one were watching and there were no external consequences, we would rape whom we wanted and steal what we needed.

Freud believed that our becoming our innermost selves is exactly *not* what the world needs. Allowing ourselves to be ruled by our innermost impulses would just lead us to hurt each other and the planet. To be yourself is to indulge yourself, and that is a frightening prospect for everyone. We should be a little bit scared of our own and each other's innermost selves.

This is the view of human nature that led Zusya's students to believe the greatest aspiration is to be more like Moses or Abraham and led Ananda to think that the True Path meant doing more of what Buddha said. We have to exert control over ourselves. To build good lives for ourselves and a peaceful world for others, we have to repress and squeeze ourselves into the molds of the saints and heroes who came before us.

There is no inherent life purpose other than survival, though it may be helpful to conjure one if it helps you rechannel your fundamental animalistic self into pursuits that are more acceptable to society and thus benefit your overall survival.

THEORY OF HUMAN PURPOSE #3: WHAT A LIFEQUESTER OR ANY GOOD HUMANIST PSYCHOLOGIST WOULD SAY

You are fundamentally good. You have everything you need within you to make you and those around you happy and safe and to make meaningful contributions to your wider community. Your growth potential is tremendous. When you peel away the layers you will find something even better and more powerful than what resides at the surface.

According to this view, our deeper authentic Selves contain so much more than simply the lust and aggression Freud believed lay below. Our organisms—as humanist psychologists often call the Self—contain tremendous creative potential and wisdom. Carl Jung, Carl Rogers, Fritz Perls, Viktor Frankl, and others founded the school of humanist psychology in response to the pessimism of behaviorist and psychoanalytic schools.

It is true, of course, that we like pleasure, yet we are not mindless pleasure machines. Yes, we can act greedy and selfish at times. But when we squish ourselves into society's single track in order to avoid being hedonists and narcissists, we have the contrary effect of shutting down our individual creative capacities for doing good. Our dark sides, meanwhile, tend to emerge unconsciously.

Our humanity—our very essence—is not something to be transcended or pushed aside or overcome. It is something to be embraced. Because as limited as we sometimes feel, as confused as we get, as many mistakes as we make, we are not bad. We are, for the most part, good. Jung and the others believed in the value and agency of individual human beings and their tremendous potential for helping themselves and their worlds.

If we expand to include the entirety of ourselves instead of shrinking ourselves because of some external mandate about who we should be and what kind of life we should pursue, then our desire for pleasure and our capacity for good will balance themselves with the rest of our human tendencies. We will not become angels or saints, but we will become people. We will be integrated.

Conscious of our more selfish impulses and in touch with the range of our innate capacities, we will find a path of our own where we will often be able to channel our psychic energy in ways that acknowledge and satisfy our hungers while helping the world. Because that is what our *organismic valuing process,* as the humanists call it, automatically causes us to do. Trusting our True Selves, we do right by the world.

This is the fun-filled path of the naughty do-gooder—the path of the lifequester. The purpose of life is to become your Self to help the world.

On his deathbed, Rabbi Zusya wasn't worried that he hadn't become as altruistic or faithful as Moses or Abraham. Zusya didn't think God would be angry that Zusya had wasted his life by not becoming like some saintly person. Zusya thought God would be disappointed because Zusya missed the opportunity to become more like *himself.*

The great mystical rabbi believed that being in sync with God's will (or what we might call the universe or society or life purpose or whatever) doesn't require us to force ourselves into that saint-shaped mold. He did not believe we needed to find a role model—a George Soros or John F. Kennedy or Moses or Jesus Christ or Gandhi or Steve Jobs or Muhammad. He believed our higher purpose is to become more like our Selves.

Suppose Zusya was right. Suppose our job is not to suppress our desires and aversions but to accept them as we expand the attention we give to the entirety of our innermost selves. Suppose it is not a matter of trying to make the clouds smaller but of paying attention to the sunlight that fills the rest of the sky. Suppose that by expanding ourselves in that way—as opposed to shrinking ourselves—we end up not only having better lives but treating the world better, as well.

Think of all the personal energy that would be liberated if you started loving the stuff you know is great inside you instead of fighting the stuff someone else has told you is bad. This is, of course, what other wise men and women have said, too. The Kingdom of God is within, say the Christians. Become your True Self, say the Buddhists.

Doesn't that mean that our biggest responsibility to ourselves and to the world is to become who we really are and then to make all of our big decisions in accordance with that Truth? And, if that's the case, wouldn't that make life a hell of a lot more exciting? I wouldn't want to miss it. Would you?

And believe it or not, this is more or less, we will soon see, what the science following on Jane's love of fiddling with a puzzle just for the hell of it begins to prove.

Want What You Really Want Tip #4: Birds of a Feather—Hang with the Right People

The research shows that materialistic people tend to have materialistic friends and family. We tend to share the values of the people we come from and the people we hang out with. Ergo, bankers hang with bankers, yoga teachers hang with yoga teachers, and activists hang with activists.

If you don't want to be a materialist, seek out people who are less materialistic and hang out with them.

Of course, it is not necessarily a good thing to hang out only with people who agree with you, but the point here is that it is hard to always hang out with people who *disagree* with you. It's hard to become a vegetarian if you spend all your time with ranchers. It's hard to start recycling if your mom constantly argues that it's not worth the effort.

People with whom we once shared values become uncomfortable when we start to change, and they discourage our changes.

So begin looking for people who make you feel safe with your becoming yourself. Spend time with people who encourage you to find your own path. Seek out people or groups who have already adopted the lifestyle changes you want to make.

It doesn't mean you have to dump all your old friends, it just means you might add some new ones in the mix who can act as home base while you explore your path.

Why Humans Do Things Just for the Hell of It

All that I've said may seem like a lot to conclude from eight monkeys who liked doing puzzles just because, in Harlow's words, it was "intrinsically rewarding." But twenty years after the monkey experiment, in 1969, a young researcher named Edward Deci at Carnegie Mellon University also doled out a bunch of mechanical puzzles, this time to human subjects.

Over the course of a series of experiments, Deci found that humans acted much like Jane and her fellow monkeys. The humans would curiously and interestedly attempt to solve a puzzle *just because*, without any promise of a reward. On the other hand, they would lose that inherent, *just because* interest in the puzzle once Deci introduced the prospect of a reward. Though they would do the puzzle for a reward, the internal *just because* spark that would have motivated them to do the puzzle anyway disappeared.

But why?

This question became the passion of the rest of Deci's long career. What makes us do things *just because*? In what direction does our internal spark lead us if it is not interfered with? What can either reinforce or divorce us from that *just because* spark? For nearly fifty years, working with his fellow researcher Richard Ryan (my friend Tim Kasser's Ph.D. supervisor) at the University of Rochester, Deci has investigated the *just because* of human nature.

Not by theorizing but through painstaking scientific experimentation—published in hundreds of scientific articles and books by Deci, Ryan, and their collaborators—proponents of what Deci calls *self-determination theory* began to prove many of the tenets of the humanistic psychology we talked about above. They showed empirically that, left to their own devices, as long as they are able to develop normally, people will not act like pleasure machines at all. Nor will they plunder and rape.

Instead, if people are not controlled or restricted by circumstance or design, they will organically attempt to fully develop their individual potentialities—become themselves—in ways that release energy and capabilities and wisdom that can make them more useful and helpful to their immediate and wider communities.

As long as they are allowed to develop naturally, people will become what they are meant for. They will become themselves to help the world and help the world to become themselves.

Deci, Ryan, and their colleagues have proved the existence of three inherent biologically coded but previously unidentified psychological needs. The behaviorists said that all human behavior was motivated by the needs for food, water, sex, sleep, and the rewards and punishments of the environment. The work of Deci and Ryan has proved there is much more to us.

Here are the things our internal spark will lead us toward if it is unimpeded:

1. Autonomy—Being Ourselves:

Deci's science shows that people have a hardwired need to feel that they are doing what is interesting to them, personally worthwhile, and vitalizing. What Harlow's and Deci's puzzle experiments demonstrated is that rewarding primates for their behavior takes away their feeling of autonomy and can demotivate them. We need to believe that we behave the way we do because we choose it rather than because we are forced by something or someone to do it.

Literally hundreds of experiments in areas from education to business to self-care to sports to personal relationships to environmental responsibility show that people left to be autonomous and approach things in their own ways and for their own reasons are more likely to bring their whole selves to the tasks, accomplish more, and be more effective and healthy.

If we feel we are following our own paths, we are turned on and awake. If we are following directions, we go back to sleep.

In part, this means that when we choose to dedicate some of our time to helping the world, we will have more energy for it if we do it autonomously—if we work on what most concerns us in the way that appeals to us. So knowing what concerns you most in the world will help inform your True Direction, whether it's lack of human connection or hungry children or animal welfare or fracking or clean drinking water or the elderly or something else.

But also, living autonomously when it comes to our life choices means we can bring more of our life energy to our communities and our world. This is because it takes so much energy to try to be straight if you are gay or to be an engineer if you are an artist or to sit in a cubicle when you are naturally active or to chase after the standard life approaches when they aren't what you really want. If you live in acceptance of your Self and allow your Self to emerge, how much more energy is then released to help others?

Maybe that is why, on his deathbed, Zusya was so upset—because he knew he'd brought less of himself to the world than he could have. True, he might have had a little bit of Moses in him to give to the world. He probably had a little bit of Abraham inside him. But if he could wake himself up, he had all of Zusya inside him to give to the world.

2. Competence–Feeling We Make a Difference:

The need to feel competent helps explain why Harlow's monkeys and Deci's humans enjoyed solving their respective puzzles. Deci and his colleagues proved that humans have an innate psychological need to feel as though there is a relationship between our behavior and a desired outcome. That we have agency. That we are able to accomplish things.

For this reason, people naturally seek out optimal challenges that explore, develop, and extend their capacities. By *optimal,* the self-determinists mean neither trivially easy nor impossibly difficult. In other words, we seek situations and tasks and projects and activities and life paths that test us but don't defeat us. Surmounting such challenges satisfies our psychological need to be competent.

The need to feel competent—and to avoid feeling incompetent—helps explain why so many of us have a tendency to ignore the problems in the world and our communities. Faced with the apparent vastness of the problems, we deny them in order to avoid the pain of feeling helpless and inept.

This is why breaking the problems down into a manageable scale that we have the ability to influence helps motivate us to actually try. This is why showing people *how* they can help is so much more motivating than chastising them about *why* they should help.

Frustration of our need to feel competent also helps explain our discontentment with the standard life approaches of the current economic system. It used to feel like the standard life approaches could bring us the security and meaning we wanted if we did the right thing. Life in the current economy feels more capricious. There no longer seems to be a clear connection between hard work and rewards, so we feel incompetent and bad about ourselves.

For the lifequester, the question becomes, how do we make *new, autonomous* life choices that give us the sense of control and agency we need? What kind of relationships should we form with the world in order for our efforts to deliver both security and meaning? How do we live our lives so we don't feel like passengers watching in horror as our captains steer our ship straight at the iceberg?

The need for competence also suggests that if we wish to take part in fixing our own lives and our worlds, we need to choose ways that play to the strengths, capacities, and talents that make us feel competent. If we want to be effective, we have to do things we are actually good at. We have to follow our passions. We need to accept and be ourselves.

3. Relatedness—Knowing We Help and Will Be Helped:

Some writers blame all the world's problems on people "being themselves." They equate autonomy with a narcissistic need for self-expression and a hedonistic obsession with desire. The truth, self-determination theorists conclude, is that the need for autonomy and competence are balanced by a third need—for relatedness. Because of this need to love and be loved, being true to our Selves means attempting to be autonomous and competent in the presence of, in service to, and with the blessing of our communities.

On the other hand, stick a person in a factory and tell him that his ideas don't matter and that the only thing he should care about is how fast he can screw in a bolt and it's true that this person may begin to act like one of Skinner's pleasure machines. Tell a person that something as fundamental to her as her sexuality is unacceptable and her unconscious will fill up with repressed urges.

Frustrate the needs for autonomy and competence and before long—it is true—people may exhibit a narcissistic preoccupation with what others think of them and a hedonistic need for pleasure. But both selfishness and narcissism are bogeymen that come up when people's needs for autonomy and competence are frustrated, not when they are indulged.

What is amazing, according to the self-determination science, is that as people become more autonomous, they actually have a greater capacity for self-regulation because there is more in their consciousness—they have repressed less. Rather than fighting their inner conflicts and demons in order to follow societal directions, they actually have the psychic space to take care of other people. They are *more* rather than less related.

A perhaps unconventional thought results from all this:

To embrace the conclusions of this scientific work presents a big challenge—especially to moralists. These results mean we must learn to trust autonomous human capacities and choices—in both ourselves and others—rather than be scared of them. They mean we must learn to trust people not just when they share our religious or political beliefs but when we sense they are truly alive and in touch with their capacities and compassion.

The difference between individuals and their trustworthiness in terms of doing right by the world is not whether they are Republican or Democrat or rich or poor or gay or straight or any of the other cultural identifiers we tend to lean on but the extent to which they are authentic and in touch with their own internal valuing process. In other words, it just may be that *the extent to which people can be trusted to care for global humanity and their community is the degree to which they are in touch with and living in line with their own individual humanity.*

A second, less unconventional but equally interesting thought:

Remember back when we were talking about the Aspiration Index and materialism, the work of my friend Professor Tim Kasser? If you remember, Tim's work showed that when you chase after materialistic or *extrinsic* goals—like financial success or popularity—you become less happy than when you embrace *intrinsic* goals—like affiliation and community feeling (see page 105 for a reminder of the different kinds of goals).

Well, it turns out that the reason intrinsic aspirations make you happier is because they help you meet the psychological needs for autonomy, competence, and relatedness. Meanwhile, materialistic goals take time and energy away from intrinsic goals, preventing us from meeting our needs.

Therefore, while following directions is not the path of the lifequester, *we are likely to find, if we follow our internal guidance systems, that our lives will naturally center on the goals of improving the world around us, having satisfying relationships, staying healthy, accepting ourselves, and staying safe and secure.*

The Psychological Organization of Goals

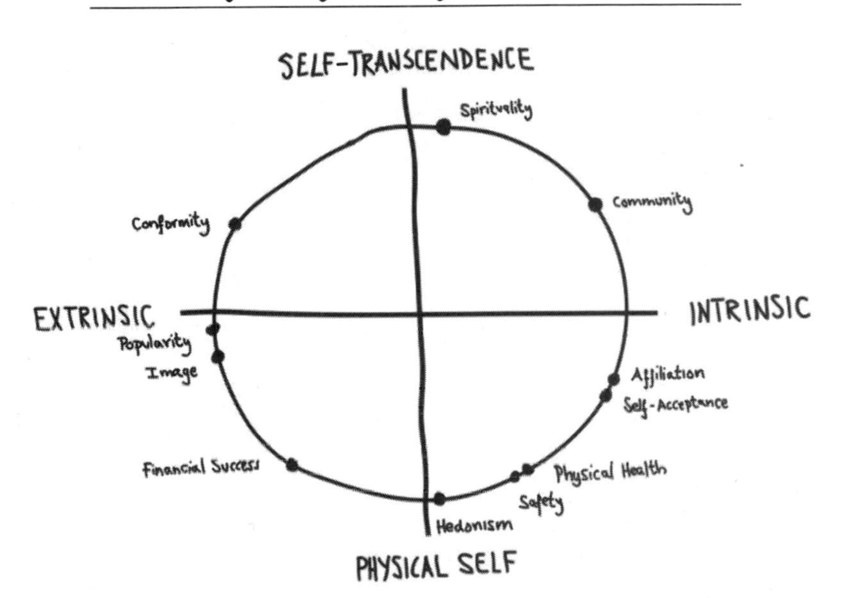

On the goals circumplex, created by Tim Kasser, Fred Grouzet, and their colleagues, goals next to each other are psychologically consistent with each other. People who work toward self-acceptance, for example, would typically also work toward affliliation. Goals on opposite sides of the circumplex are in conflict with each other. People who emphasize financial success, for example, tend to put less weight on community feeling.

How to Be an Everyday Hero

In 1949, when the scholar Joseph Campbell published *The Hero with a Thousand Faces,* he told us that every culture in every time throughout history has or had myths that coalesce around a basic theme—the hero story.

As the hero story typically goes, a societal misfit of some kind cannot conform to his community's standard life approaches. He leaves or is somehow forced to depart on some sort of adventure. During the adventure, the misfit discovers that the very characteristics that made him an outsider can actually provide tremendous strength if they are developed and directed rather than repressed.

After finding himself uniquely suited to surmount many challenges because of his uniqueness, the misfit can now return to the community, no longer someone who cannot follow his societal rules but someone with the confidence to teach and lead people in breaking those rules—either quietly or boldly. What once alienated the misfit is now a gift that provides him with the power to help his community, or at least certain members of his community.

This story is repeated over and over in our culture: the obese person who loses weight and helps other people do so, the lonely Wall Street banker who becomes a yoga teacher and leads others down the path to fulfillment, the breast cancer sufferer who survives and starts a cancer nonprofit, and on and on.

The hero story does not have to be so grand. It's just the story of uncovering a part of yourself that you previously rejected, bringing it into the light, developing it, and then using the new strength to help yourself and others. Actually, most of us—at least those of us who choose to grow and change—cycle through the hero story hundreds of times in our lives in different ways. To be an everyday hero, in some ways, is to simply fulfill the psychological needs for autonomy, effectiveness, and relatedness we have discussed.

The word *hero* can feel intimidating because it has been conflated in our culture with *superhero.* But being an everyday hero is just a way of becoming yourself that contributes to your own personal world and helping your own personal world in a way that helps you become yourself. It is simply the path of the lifequester.

A Bike-Riding, Dumpster-Diving Former Teacher Finds a Noncareer, Community-Based Path to Security and Meaning for Herself and Others

What follows is the story of my friend Rebekah Schiller, who embraced her differences and found a way to quietly use them to help people. That is what makes her an everyday hero, a lifequester.

I remember feeling alienated as far back as when I was twelve years old, going to a private school in Alabama. The kids all came from well-to-do families and I didn't fit in. We had to wear uniforms, so to be myself, I wore rings on all my fingers and crazy fingernail polish and colored my shoes. I hung out with the kids in the anarchist and humane societies. It was all very white and privileged and felt separate from the real world that most people live in and that I wanted to be part of.

My parents had this idea that I would study art in Paris and they would visit me, but I loved science. I love knowing how everything works and figuring out how to fix it. So after I graduated from high school magna cum laude and went to college, I ended up at Stony Brook University on Long Island pursuing a Ph.D. in physics.

I spent eleven hours a day in a dark lab with no windows and I loved, loved, loved it. But again, it started to bother me that I lived in this land of privilege where we were apart from the rest of the world and where no one seemed to really worry about that world. I began to feel alienated again because no one really shared my concerns.

The feelings of separateness got to be too much, and I quit to join the NYC Teaching Fellows program, where I thought I could have a more direct impact on people's lives. The city pays for your master's in education while you actually begin teaching. I ended up at the Bronx High School of Science, a public school with a high-stakes academic environment. But I began to worry that the school was more concerned with maintaining its standards than helping the kids.

I had one student, for example, who had such an awful home life, and he said the only reason he came to school was because he wanted to take my class. But he was underperforming because of the trauma at home and an administrator told me that they would have to remove him from my class. The student begged me. He said, "I can't go to school if I don't take this class."

I started to question the administration's view about why we teach physics and why kids go to school and what education is about. If your home life is a mess, then learning physics could be the perfect thing for you. It helps you model the world and learn to problem-solve. Using your brain in that way is the point of learning, not jumping through academic hoops and getting good grades.

I love teaching. I love high school students. But I started to feel I couldn't work in a school that wasn't as much about the teaching or the students as I wanted it to be. I felt alienated again. I decided I would save $30,000 and then I would quit and figure out a path where I could find what I wanted from my life. Back then, I was still thinking that finding the right career was the magic bullet.

Meanwhile, I had become involved in Grub Community

Dinners here in Brooklyn. Grub is a regular "freegan" feast that is free and open to anyone. Most of the food is rescued from stores that are trashing the food as it approaches its sell-by date. So Grub serves the purposes of providing community and feeding people while demonstrating a model for reducing societal waste through sharing.

At Grub, I found a group of people who seemed to really care what was happening in the world and were actually experimenting with ways of living differently. Going there helped relieve my sense of alienation. This was back in April 2011.

When Occupy Wall Street (OWS) started in September 2011, some of my friends from the Grub community were among the first at the protests. I didn't go initially. I was cynical about the power of protest. Then we decided that we would hold one of our Grub dinners at Zuccotti Park to help feed the Occupiers. We brought the meal by bike from Brooklyn. That was my first time there and I was really impressed by the OWS community in the same way I had been at Grub.

I started going to the park once a week. I was very inspired. I had never really thought about structures of government and decision-making processes. The methods of developing consensus at OWS were very striking to me. Everybody, no matter who they were, got to express their opinions, and all opinions were synthesized in an agreement that included all views. It wasn't about the majority simply outweighing the minority. I really felt at home.

Around that time, I learned about the Icarus Project, a radical peer-support group for people diagnosed with mental disorders. Two people who had been diagnosed as bipolar were the founders and I liked the language they used. They said it is not crazy to be crazy in this crazy world. I wasn't involved with them, but their thinking appealed to me. If you don't fit in, maybe it's not your fault. Maybe it is the world's fault. After all the alienation I felt, I identified with that.

That idea, and finding the Grub and OWS communities, helped me realize I might not be crazy to want to quit my teaching job at Bronx Science. It solidified that feeling that I don't need to work in a system if I don't agree with it. I didn't need to change myself to fit my life but the other way around. I needed to change my life to fit myself. Plus, I was witnessing from my friends that you don't need that much money to thrive if you have a strong community.

I was invited to join a cooperative living situation for around fifteen people. We ate and cooked and ran the house together. We all wanted to live with people who work on similar community projects. We all had the view that there is something about the world that needs to be changed. We wanted to provide concrete alternatives—the way that I saw Grub doing—and help bring attention to them.

Grub was a project of a group called A New World in Our Hearts. It also had a project called the Brooklyn Free Store, a pop-up sharing space where people could take things they had a use for and leave things they wanted to share.

We went to colleges and got the stuff the kids were all throwing away as they left for the summer. The Free Store ended up with shelves of books and racks of clothing and tables of housewares. It is a model for another way of living without money and without scarcity. People meet each other and learn to get security from community instead of the corporations they may feel they have to work for.

Meanwhile, I was dealing with feeling traumatized by the police violence I witnessed at Occupy Wall Street. Before Occupy, I never paid attention to the justice system. As a middle-class white woman, I thought I would never get arrested as long as I didn't commit a crime. But that's not what I saw at Occupy. Instead I saw people peacefully exercising their First Amendment rights being beaten and arrested by police.

I began looking for alternatives to what we do to each other in that kind of justice system. I trained to be a community mediator with the New York Peace Institute. We work to help resolve family and community conflicts peacefully through negotiation instead of trial and lawsuit. Our clients talk together about what they need to work out and they decide together how they want to move forward.

Ultimately, I hope to begin earning my money through my mediation work, but I've discovered I don't need much cash. Living in community, my rent is only $400 a month. I ride a bike everywhere. I Dumpster-dive and am part of a community where we share. All my clothes are from the Free Store. I love my life and it would be impossible without friends who also were committed to sharing. It would also be impossible if I had a full-time job. I get by on what I earn from one day of teaching a week. Last year I lived on $7,000.

I used to want to be financially self-sufficient—that was why I thought I needed that $30,000. But one thing I've learned is that with money you can so easily buy isolation. You can sit at home and get food delivered to you and, if you use a credit card, you don't even have to pay the person who delivers it or look them in the eye. You can fill up your car with gas and you don't have to think about where it comes from. On the other hand, when you don't have money you start to be more aware of the interdependence of people.

Maybe interdependence is what I was always looking for and what I always wanted to help other people find. I really found the group of people who share values with me. You don't need to help each other when you have lots of money. I don't want that isolated way of being. I love my life. And I am completely confident that I'm doing exactly what I want to do.

Want What You Really Want Tip #5: Get Some Happiness

It turns out that one way to be less materialistic, to feel free to become yourself, to have the energy to help others, and to want what you really want is to be happy. Think about it: if you feel happy, then you worry less about what you don't have and have more freedom to do and be what feels right to you. You feel less insecure and more centered in who you are.

Happiness for oneself alone is not the end goal for the lifequester. But it does turn out that being happy provides a certain kind of energy that lifequesters can use to change their lives and their world. The research shows that happier people are less lonely, less materialistic, and naturally more helpful and empathic toward others. A depressed lifequester sits on the couch; a happier lifequester does things. Happiness is a kind of mental fuel that helps the lifequester pursue Truth.

For these reasons, as we learn to be ourselves and to follow Truth, it can help to borrow from the methods of positive psychology. The following methods have all been scientifically proven by various studies to improve happiness (I draw most particularly from Sonja Lyubomirsky's work):

→ **CULTIVATE GRATITUDE:** One method is to write down and think about five things you are grateful for, just once a week. Another is to write letters of gratitude to important people in your life.

→ **CONNECT IN PERSON:** Of course, spend more time with people you like and love, but Lyubomirsky also suggests acts of kindness. She recommends one large act of kindness that is outside your routine each week and to vary it so it feels novel and not like a burden or an obligation.

→ **BE MORE PLAYFUL:** There is a state of mind that scientists call *flow,* where we disappear into an experience and self-consciousness is extinguished. It comes in different ways for different people—laughing with friends, throwing a Frisbee, cooking a great meal, etc. So do more of whatever feels like play to you.

→ EXERCISE MORE: Studies show that people who walk only 120 minutes a week—20 minutes a day for six days—were 63 percent less likely to suffer from depression.

→ MEDITATE: Meditation trains you to be more detached from the negative judgments that cross your mind and to be able to act more from your center. For a detailed discussion of and instructions for meditation, see page 425.

9

WHY MOST OF HUMANITY SEEMS TO BE HEADED STRAIGHT TOWARD THOSE DEATHBED REGRETS AND HOW TO CHANGE COURSE

How a Dog Loses His Dogness

Here is a true story about a dog who lived in a cage with a metal bottom. Let's call him Ralph. One day, back in the 1960s, a now-well-known positive psychologist named Martin Seligman applied a small electric current to the cage floor—just enough to make it uncomfortable for poor Ralph to stand there.

At first, Ralph ran around looking for a place to stand that wasn't painful. Sadly, there was no such place. Thankfully, though, the electricity eventually went off, but not because of anything Ralph had done. Nothing Ralph tried had actually mattered.

Poor Ralph the dog lived through this a few more times, experiencing over and over the futility of his actions. The electricity tingled through his feet or it didn't. The pain came and went. But over and over again, Ralph found that his attempts to change his situation didn't help or make a difference.

Eventually, Ralph resigned himself to the discomfort and stopped trying to get away from it. When the electricity came on, he just sat there. Sometimes Ralph whined. Sometimes he shuddered. Sometimes he peed. But before long, Ralph acted as though he had pretty much decided there was no point trying to escape and he gave up.

Meanwhile, another dog lived in another cage. The floor of this cage was electrified on one side but not on the other. When Seligman turned on the juice, this other dog ran around and quickly realized that he could jump over a small barrier into the part of the cage that was not electrified. It took the dog a couple of seconds to figure it out.

Soon enough, poor old Ralph ended up in this new cage. Again, one side was electrified but, just a short distance away, the other side was not. When Seligman turned on the juice, however, Ralph just stood there. The other dog had relied on his inner resources and learned that escape from the electricity was literally three steps and a little hop away.

Ralph didn't even try. He had generalized from his experience with electric shocks in one situation the incorrect conclusion that his own efforts and ability to figure things out and escape were futile in all situations. He had acquired a condition known as "learned helplessness."

You could say Ralph became alienated from his own potential. He no longer trusted that his innate abilities could help him. He kind of gave up. He lost some of his dogness—his inner spark, as it were.

It wasn't that he wouldn't eat food or drink water when it came. It wasn't that he didn't want to stay alive. But when it came to avoiding shocks and discomfort and improving his lot, he no longer bothered to try.

How a Person Loses His Personness

Now, imagine a *person* named Ralph. He hasn't stopped going to work and earning money and eating and drinking and even having sex and going on vacation when he gets the chance.

But some time long ago, maybe in his childhood, one shock too many taught Ralph the person that, at least in some areas of life, he is kind of helpless. Ralph the person has generalized some childhood lessons that certain kinds of goals just can't be achieved and certain needs just can't be met.

Let's say Ralph started out, in his youth, naturally loving to dance. It wasn't necessarily anything as self-conscious as a career choice. It was just his passion. It made him feel good. He danced with other people and he was good at it and so it filled his psychological needs to feel autonomous and competent and related because it was what came naturally to him.

But let's say Ralph's dad drank too much and was liberal with his fists. Or that there wasn't enough money and his family got evicted a lot. Or that his parents argued too much and got divorced. Or that there were earthquakes in his city or a war in his country. Or even that people rejected him because of his love of dance.

Basically, let's say that any number of things happened that made Ralph often feel insecure or unsafe or frightened. Psychologically, it began to feel too dangerous to give his attention to his dance because he was too busy avoiding punches or worried about his mom.

In the area of achieving certain goals and satisfying needs like those for autonomy, competence, and relatedness, he kind of coasts. He puts those needs aside. He thinks, "There are more important things to do than dance. Like find a way to be safe. I don't need relatedness. I need money for a big house where no one can interfere with me. After that, maybe I will dance."

The research does show that people who grow up in insecure circumstances like Ralph's turn to materialistic goals. Their attempts to achieve intrinsic goals and meet their psychological needs often get frustrated by the circumstances and they start to learn to be helpless in those areas. Of course, there are other ways people become materialistic, too—like just learning it from their families.

But what happened to Ralph was that he was insecure and couldn't tolerate that feeling, and he decided to put all of his effort into his excessive materialistic ambitions—the best job, the biggest house, and all the other consolation prizes that at least seem attainable—to stop the feelings of insecurity. He stops dancing and goes to school to become an accountant.

Admittedly, he doesn't much like accountancy, but he trundles along, because that's what you have to do, right? Follow directions.

As the years pass, Ralph grabs at life's carrots and runs away from its sticks. He has built a life based on achieving and earning, and now the prospect of stopping all that is hugely frightening. He feels stuck. What if he can't pay the mortgage? What if he can't meet his credit card bills? Everything is based on the narrow path he has chosen, and he sees no way to veer off it.

He has no choice but to ignore whatever it was inside him that first led him to dance, and with time, he loses contact with his fundamental spirit altogether. If you ask him about his True Self, about the direction of his life, he has trouble understanding what you mean. Instead, he prefers to give his attention to what he knows he can attain.

When he is stuck in a traffic jam for the fifth time in a week, he doesn't think, *What am I doing living a life full of traffic jams?* Or *What are all these stationary cars doing here, pumping out all those greenhouse gases?* He doesn't question. He has turned that part of himself off.

Instead, he looks at the car next to his, blaring out music, and thinks, *I want a Mercedes with a Bose stereo like that one. I'd feel better then.* Or maybe he daydreams and thinks, *I want an Eames chair in my office.*

Meanwhile, he is too tired to do anything but watch TV when he gets home, and the media culture keeps hitting him over the head with images of people who have and do more than he does. Advertisements show people with better things and more friends and louder laughter. The images make him feel insecure and scared that he is missing out on life and further activate his materialistic goals—exactly what the ads are designed to do.

Ralph laughs out loud when people remind him of how he used to love to dance. He doesn't even remember anymore what it feels like to have free time to spend with people he cares about and do things he loves. He doesn't think to learn the guitar or to make a huge sculpture or to play with the kids who live on his block or to go for a run or to help save the Pacific dolphins from fishing nets.

Those impulses have all been frustrated in the past, and so he is a bit like Ralph the dog after he is put in the second cage. There is a chance of escape, but instead of looking for it, he just stands there when the electricity is turned on.

He sees those advertising images of those happy people and wants to feel the way they seem to feel and figures that to get those feelings, he needs to do what the ads say those people did. Like get the nice sound system in his car or a fancy office chair. These are things he believes he can actually do.

Like Ralph the dog, Ralph the person has stopped trusting his innate abilities. Or his talents. Or his spirit. He is barely even aware anymore of his internal guidance system and what it is telling him. He doesn't know why he can't find the joy in his life or why he always feels a pressure in his chest like he should be something more.

He ignores the news about climate change and other world crises because he has no idea how to act outside his narrow parameters. How is anyone supposed to do anything about that stuff? Plus, his own life feels so precarious and unsatisfying. How is he supposed to take care of other people when things just feel so bad for him?

Instead, he puts more and more effort into following his employer's and society's directions—work harder, look better, buy more stylishly—but he still feels empty. Because those intrinsic psychological needs that were filled so well when once Ralph danced have never gone away. He has learned to ignore them, but still they nag.

Ralph tries harder and harder at the wrong things to fill the hole in his soul, but they don't work. The more alienated he becomes from his life force, the more he calcifies into a rigid, living automaton. If there is a reward to be had—more money, say—Ralph might give it a halfhearted effort. If there is a punishment to be avoided—more work, say—Ralph will look for the shortcut.

He has become like one of B. F. Skinner's pleasure machines

and so, without knowing it, and in some ways without any blame, kind of on autopilot, Ralph and so many others of us come to head toward those deathbed regrets. This is how some of us have lost our personness and learned to turn the volume down on who we really are.

One little tangential thought before we move on:

If you care about the world or have started to peek your head out of the Matrix, you may be in a lot of despair about the state we're in. You may look around and see lots of Ralphs driving their SUVs and feel angry.

That is natural.

But consider the above story again. All that our anger will do to a Ralph is make him more scared. It will make a Ralph push harder against the door that never opens.

If you want to really change Ralph, you know what may begin to help him feel safe enough to change? You know what you have to give him?

1. Your love
2. Your compassion

How to Become a Person Again

Three pieces of good news:

1. Becoming Ralph is not inevitable: When Martin Seligman did his experiment with Ralph the dog, and other dogs after him, it is true that many dogs gave up when it came to trying to escape the tingling electricity. But every so often, no matter how many times a particular dog was shocked, that particular dog kept trying. In the first cage, he or she kept running around every single time the electricity came on. Then, when that dog got to the second cage, it quickly jumped over the barrier to safety. Such dogs just wouldn't be divorced from their dogness. Becoming Ralph is not inevitable.

2. Even if you've become Ralph, you don't have to stay Ralph: When you look around the human world, you might think that so many people you see are a kind of Ralph, but look again. All sorts of people are doing the about-face from being an accountant or changing their life or suddenly deciding to be everyday heroes. Look around in your own life. Find the undramatic kindnesses. The peopleness of people is everywhere when you look for it. If you've become Ralph, staying Ralph is not necessary.

3. To not be Ralph, all you have to do is not be Ralph: In Asia, people sometimes keep on their desk something called a dharma doll. Dharma dolls have round, weighted bottoms so that when you knock them over, they right themselves again. The purpose of the doll is to remind us of our fundamental human spirit. They remind us that even in the face of our failings and difficult circumstances, we can still be fully alive and we can still try. That even if we get knocked down by life seven times, we can still get up eight. It's our choice.

A TRUTH:
All it really takes to return to our Fundamental Spirit is to consistently remember that it is there.

A SECOND, CLOSELY ASSOCIATED TRUTH:
All it really takes to help other people return to their Fundamental Spirit is to consistently remember that it is there.

We are born with gigantic potential to help both our own lives and the lives of those around us. What holds all this potential is our True Self or our psyche or our organism or whatever you want to call it. When all that potential is available to us, when we are not scared of our entire capacity, when we trust our inner valuing process, we are what one of the founders of humanist psychology, Carl Rogers, called a *fully functioning person*.

When we are born, we start out on a path toward being fully functioning, but as we move through childhood, our parents shout, our teachers give us detentions, and it becomes clear that certain behaviors will bring us love and food and protection and others will bring scorn and punishment and insecurity. Slowly, we close down those parts of ourselves that the environment punishes while exhibiting over and over, like a broken record, those behaviors our environment rewards.

We develop what Carl Rogers called a *concept of self*, a much-limited version of our True Self that contains only the set of behaviors that it is safe and acceptable and effective to display in our childhood environment. We could call this process "becoming Ralph."

To become a fully functioning person means to reclaim the parts of yourself you have closed off. To return to your full and True Self. For Ananda, that means allowing Ananda to emerge instead of trying to be Buddha. For Zusya, that means allowing Zusya to emerge

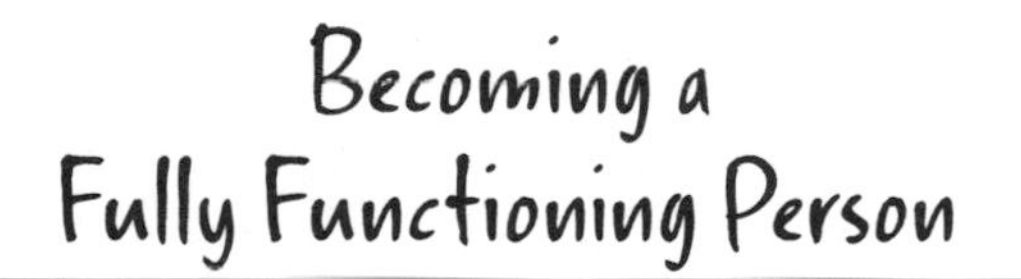

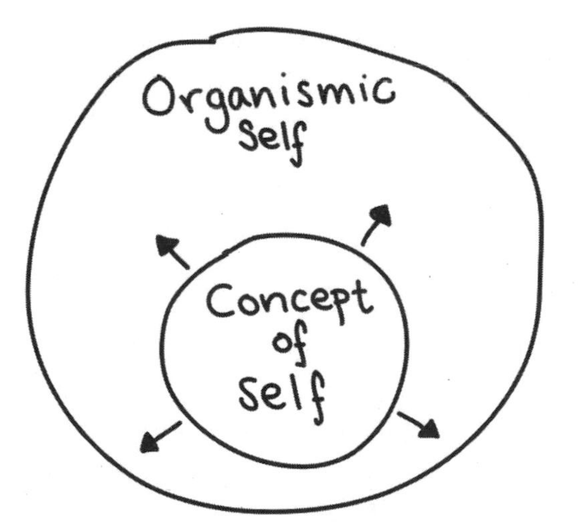

instead of trying to be Moses. For the lifequester, it means allowing life to unfold individually instead of following all the standard life approaches.

To be our Selves does not mean to be a dancer or a carpenter or a mathematician or to find the right career. Making career the be-all and end-all is just another standard life approach. Being our Selves is not about our careers so much as our ability to value and react to our experiences authentically as we go along, moment to moment.

Carl Rogers wrote:

> [The good life] is not, in my estimation, a state of virtue, or contentment, or nirvana, or happiness. It is not a condition in which the individual is adjusted, or fulfilled, or actualized . . . The good life is a process, not a state of being. It is a direction, not a destination. . . .
>
> The good life, from the point of view of my experience, is the process of movement in a direction which the human organism selects when it is inwardly free to move in any direction, and the general qualities of this selected direction appear to have a certain universality.

The article of faith is that if we achieve this Good Life, we will turn, with each step, in the right direction and find ourselves in the right places for ourselves—even if we didn't even know we were aiming to get to them—and that when we are no longer in the right places, we will move along. That our organism or Self—even if it is not all conscious—has everything it needs to help itself and help the world.

A fully functioning person will not mindlessly chase after the materialistic consolation prizes—the extrinsic goals—but will automatically and even without conscious decisions pursue intrinsic goals, just as a plant automatically grows toward the light. What stands between us and our ability to be everyday heroes or fully functioning people are the knots and blocks. Part of the journey of the lifequester is to expand beyond those blocks.

Because of the same principle of growth that causes an acorn to become an oak tree, this expansion will happen naturally if we simply allow it to. All we need to do is trust it and hear it and emphasize it more than we do the blocks. We need to create conditions where we can listen deeply to ourselves, and we need to let other trusted people listen deeply to us and witness as we experiment with new behaviors (and in return we will find ourselves naturally doing the same for others).

Use the tips in this book and elsewhere to be kind to yourself when you experience the fear of becoming more. Get support. Make yourself safe. Nurture what grows. Not for yourself but for the world. Because part of what is hidden from us is our own empathy and compassion and our own particular ways to act in the face of that empathy and compassion.

Listen to yourself. Be an everyday hero. Believe that, underneath it all, this is what you really want and value in your life. Want it because you know, deep inside, that even when the shiny things glitter and distract, becoming your full person is what you really wanted all along.

Want What You Really Want Tip #6: Talk to Someone Who Will Really Listen (and Not Judge)

Part of what helps us let our hidden capacities reemerge is tentatively expressing and discussing them in the presence of someone who accepts and values them. As we experience the acceptance of hidden parts of ourselves, we feel free to let more and more come out. We feel safe to want what we really want.

In fact, research by Tim Kasser and his colleagues suggests that being in the presence of "a person who clearly likes you, tends to be very accepting and non-evaluative of you, and simply accepts you for who you are" causes a shift away from extrinsic values and toward intrinsic values. Just thinking about such a person, the research shows, helps us live in our intrinsic values.

One way to formalize this kind of experience can be to find a good therapist who is capable of strong "unconditional positive regard." But you can also set up a co-counseling situation with a friend you really trust, where you take turns listening to each other in an atmosphere of noncontingent acceptance.

10

A LAST WORD ON WHY THE WORLD WOULD BE SAFER AND HAPPIER IF WE WANTED WHAT WE REALLY WANT

When we lose ourselves and turn to materialistic ends, we chase after money, possessions, fame, and power. Not only do we mistakenly sacrifice our happiness and well-being, but, chasing after these things, we lose our ability to hear the quiet voice within us that thirsts for relatedness and aims us toward a feeling of community.

We ignore the facts when we hear, for example, that the construction of a new sports stadium is destroying a neighborhood. Or that bush wars are being fought to get minerals for our next cell phone. Or that the government is buying more drones while closing more schools. Or that the constant quest for more is destroying both our habitat's ability to support us and our neighbors' ability to live safely.

However, a fully functioning human being—a lifequester—not only is happier and more fulfilled but values his or her life choices and community involvement differently, in ways that cause less harm, do more good, and inspire others to cause less harm and do more good.

No person—and especially not the lifequester—can be reduced to a list of characteristics, of course. But it helps to remember, as we have shown, that the scientists have proved that among the qualities universal to the lifequester's Good Life are:

Security: safety from physical harm and ability to survive.

Autonomy: the feeling that one's choices come from within.

Competence: satisfaction that one has skills and abilities that make a difference.

Relatedness: connection to others in meaningful and helpful ways.

In the next part of the book, we will look at how to choose relationships with the world that are informed by our values and what we really want from those relationships—the need to help and be helped.

Meanwhile, the most amazing news is that when you become yourself, you respond with the utmost appropriateness to the world around you. You stop fighting what is inside you and start trusting your real needs and ways of doing things.

When you want what you really want, you know that your needs cannot be met by things that can be bought for cash or molded from petroleum or shipped from China. What you want are things that can be achieved only when we strive to use our most prized capacities in service of our most fervent community concerns.

In other words, if we all wanted what we really wanted, then we would not be making ourselves unhappy while our desires destroyed the world through consumption and the constant fight for resources. We would be living the Good Life as our aspirations helped the world—following the principle of becoming yourself to help the world and helping the world to become yourself.

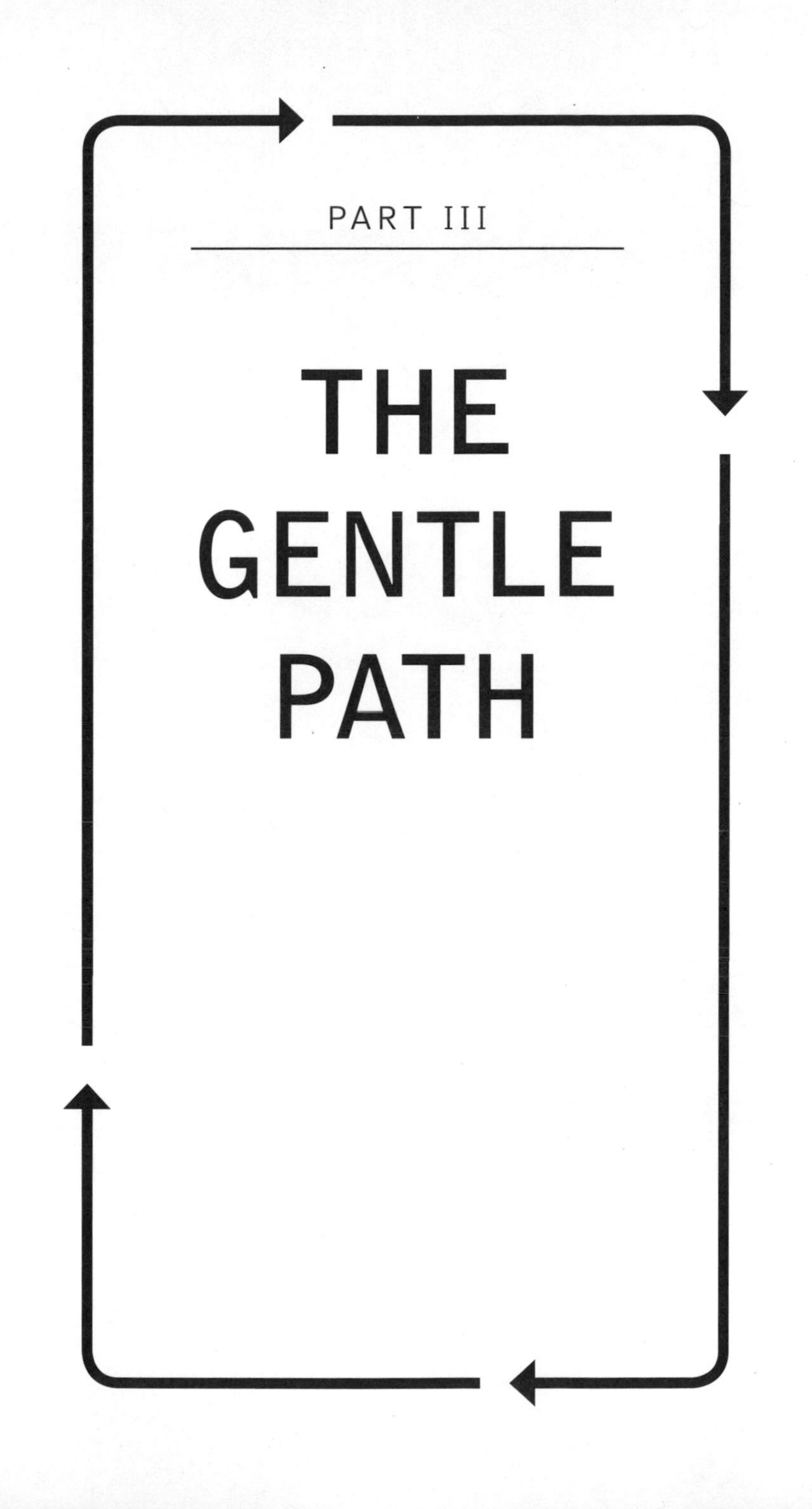

PART III

THE GENTLE PATH

11

A JOURNEY OF A THOUSAND MILES, ETC., ETC.

Following Truth by Changing the Easy Parts First

All that we've said about following our Truth is very inspiring, but it doesn't change the realities of having to earn a living, pay the rent or mortgage, feed the kids, repay student loans, care for parents, etc. When you think about those realities, becoming your True Self sounds like a lovely fantasy that can make you feel good for a little while but then overwhelms you.

That's because we so often think in terms of changing the big stuff—career, community, where you live. How are you supposed to move those mountains? They represent the very core of how so many of us live our lives. I call that approach "change from the inside out"—from the core of your life to the fringes—and it *can* work for some people.

But a gentler approach means giving our attention to any one of the much smaller relationships with the world. I call that "change from the outside in."

So give yourself a pass on the big stuff (for now). Find a way to truly be your Self—to hold to your values—in any one of the much smaller, easier-to-wrap-your-head-around relationships you have with the world. Like with food. Or stuff. Or community.

Over time, and treating yourself gently and kindly, repeat, repeat, repeat with the next and the next and the next small relationship. Fixing one tiny relationship at a time, you make your life a little better and the world a little better, too. Bit by bit, you will find yourself spending more time with people you enjoy, doing more things you value, helping the world in more ways you care about. You'll even find that your small changes from the outside add up to a life that supports making the big changes—if you want them—and makes them seem manageable.

That's working from the outside in. Suddenly, moving toward that authentic, meaningful Good Life doesn't seem so hard, right?

Four More Reasons to Put the Easy Things First and Worry About Career Last

1. COOKIE-CUTTER CAREERS ARE NOT THE EASIEST PLACE TO FIND MEANING

Oftentimes, in our culture, when we talk about finding the Good Life, we automatically assume that we are talking about finding the best career, as though our career is our entire life. But as we have said, changing career is an onerous first step. It is big and scary and hard. So it might not be the right place to start. Additionally, if finding meaning and purpose comprises the Good Life, then our careers, according to the data, really may not be the best place to look.

To wit:

> Only 30 percent of workers in the United States feel engaged by their jobs, according to a 2013 Gallup report.
>
> Only 18 percent of workers have time to think creatively.
>
> Only 38 percent feel they are growing personally.

> Only 25 percent feel connected to their company's mission, according to a study of 12,000 mostly white-collar workers published by the *Harvard Business Review.*

After all, staring at a screen at an insurance company is staring at a screen at a law office is staring at a screen at a bank is still staring at a screen in a cubicle. We have to be careful that we don't, by joining in society's massive game of musical cubicles, miss the *real* opportunities to build good and meaningful lives.

What if you put your effort into making more friends or seeing the ones you have more often? What if you learned how to play music? What if you learned how to cook? What if you joined an environmental group? What if you found the one easy thing outside of career that would bring more meaning and purpose to your life?

The standard life approach is to obsess over career to the point that we overlook other elements of our lives that might actually have better or at least more immediate potential to bring happiness to ourselves and our world. Many of the lifequesters I've met, in their frustration with those screens and cubicles, take entirely the opposite approach.

Lifequesters take the time and energy to gently build their own personal good and meaningful lives outside of work. Then, instead of finding a career and fitting their life to it, they find a career—or better yet, a vocation—that complements their own personal Good Life.

2. IT'S HARD TO FIND YOUR REAL JOB BEFORE YOU KNOW YOUR REAL SELF

How much free time do you need? What community should you live in? Whom should you live with? What talents do you have? Should you be married? Have children? Be single? Play the field? Are you gay? Are you straight? What concerns you most in the world? Who *are* you?

Most of us starting out on our lifequests are so used to the standard life approaches that we can't even imagine what our Good

Life looks like. That's partly because the standard approach doesn't encourage us to prioritize figuring out who we really are.

By experimenting with our smaller relationships with the world—how we eat, what we own, whom we socialize with, and all the other stuff—we learn what we need to know about ourselves to build our personal Good Life and, later, to discern what vocation we are called to. Find the real you and then you can find your real job.

3. BUILD A GOOD LIFE NOW AND CAREER FREEDOM COMES LATER

The standard life approach is to put career first and then try to fit a decent life around it. But when we take that approach, we have to accept some choices that seem almost crazy:

> We move away from the people we love so that we can get a high-paying job that will earn us enough money to—you guessed it—visit the people we love.
>
> We spend our days sitting in front of computers to earn money to pay a trainer to help us work off the gut we grew because all we do is sit in front of computers.
>
> We work overtime to pay for the babysitter who looks after our kids while we work overtime.

Some of us, before we know who we really are, follow the societal herd. We buy the idea that we need a lot of money without knowing what we need the money for. Like somehow the money itself will fix things. So we join the career merry-go-round and suddenly we are in this crazy cycle that we feel we can't escape because we think we need the money. We have sacrificed our freedom.

What if we figure out how to live our own version of the Good Life, one that fills our real needs—to feel secure and healthy, to be ourselves, to relate to others meaningfully, and to use our talents? What if we find as much security and meaning as we can by building good lives outside our careers? Because if the rest of your life is wonderful, then how much money and how great a job do you actually need?

With a Good Life as a foundation, you don't have to have the perfect income. You don't have to find your entire purpose in your work. Suddenly there is less pressure on your career choices. That's career freedom. With a Good Life as your foundation, you are more likely to be able to choose or change to the career you want instead of the career you need.

4. EVEN IF YOU LOVE YOUR CAREER, WILL IT LOVE YOU BACK?

Forty million full-time U.S. workers are no longer permanently employed. Instead, we are independent workers such as temps, contractors, and freelancers. Meanwhile, four out of five American adults will face unemployment at some time in their lives. That puts us on a constant where-do-I-get-my-income-and-health-care-and-retirement treadmill.

A "good" career is not a guarantee of security anymore.

When lifequesters work on their non-career-related world relationships, though, many of them stumble on new ways to find security and happiness. Maybe finding themselves part of a new community alleviates the need to pay for babysitters. Or a passion for vintage clothing brings wardrobe costs way down. Or emancipating oneself from the automobile saves the $10,000 on average spent annually on car ownership.

In building their Good Lives, many lifequesters find themselves sharing, trading, bartering, and cooperatively producing everything from electricity to food to living spaces. They use at least some of their time to produce value directly, rather than going to work to earn money to buy that same value.

Most of them find that this approach makes both them and their worlds happier as they nurture the full range of their relationships rather than just the one with their employer. They help others as they help themselves. They get security and meaning. Sociologists even have a name for it: "diversifying into the nonmarket economy."

Or we could call it "loving what has the capacity to actually love you back."

But Wait! I Still Want a Great Career!

None of this is to say it isn't possible to have a great career. I certainly love mine, and I know many other lifequesters who love theirs, too.

But part of what makes a career—actually, a vocation—great by lifequester standards is that it fits one's own version of the Good Life and is true to the Self. What is great about a lifequester's vocation is not how "successful" it is but that it fits with the rest of what the lifequester is passionate about—including spending ample time on other things and with the people he or she loves.

The point is that *exclusively* chasing after a secure, meaningful career is a limited objective. Besides, you may already earn well and get a lot of satisfaction from your present career (or feel confident that you will get these things from the career you aspire to). That is great.

But why not aim to get security and meaning from our *whole lives*, in *all* our relationships with the world? For those of us struggling to get a foothold on the career ladder, this presents an alternative strategy. For those of us with excellent careers already, it is the next frontier. This is another reason to make a gentler start and to begin by attending to some of the smaller relationships with the world.

12

IDENTIFYING THE EASY PARTS

What Things Do You Wish Over and Over Again Were Different?

Jonathan wanted to play music all his life and always imagined guitar would be his instrument. He fantasized about sitting with other musician friends on sunny afternoons under a tree in Brooklyn's Prospect Park strumming and singing. But he was stuck in a busy job and life. He thought playing music—an important part of living a life in line with his values—was just a pipe dream.

"I always thought, what is the point of even trying?" he said. "Guitar takes so long to learn and there are so many amazing guitarists in the world. Who has the time to get good enough that other people want to play with you?"

Sound familiar?

Do you use your time and money in ways that actually make you and your world happy?

This is a theme I raise with my speaking audiences around the world. I say, for example, "Raise your hand if you play a musical instrument." Usually, about five out of a hundred raise their hands. Then I say, "Raise your hand if you wish you played a musical instrument," and then closer to half the audience raises their hands.

I usually respond, "What is the point of even being alive if you can't play guitar when you want to?"

People laugh, of course, but only because I'm making a joke about something all of us feel—that there is no room in our lives for doing the things we actually value. We don't just feel that we can't find room for music. We feel we can't find room to eat healthily, exercise, have romance and sex, make art, spend time with our kids and loved ones, help out in our communities, or address our world concerns.

Part of that is societal. Time-use researchers report that Americans spend decreasing hours in the workplace but still have less time to have fun. Because when we aren't working our butts off, we are driving our kids around or commuting for hours or answering our colleagues' e-mails from home. We are in the middle of a time drought, caused in part by having lifestyles that kind of suck.

But part of it, too, is because we are trying to change the big stuff. We keep facing off against the biggest challenges and then giving up. What if instead of setting up a challenge to "eat differently," we just, today, had a slightly better lunch? What if instead of trying to find "more time with the kids," we just, today, got everyone to the breakfast table fifteen minutes earlier and kept all gadgets off? What if instead of "doing something about" climate change or institutional racism, we just go on that one march?

There may be huge reasons why we can't make the big changes we'd like to make in our lives. But most of life is actually lived in the small decisions. So let's accept and work with our limitations rather than being stopped by them. Let's turn the problem on its head and ask: Given that we have so many time obligations, *how do we still make even the tiniest starts in living the Good Lives we want?*

Identifying Easy Parts #1:

Right now. Take five minutes and make a list on an envelope of improvements you'd like to make in your life. The longer the list, the better. Now put parentheses around the really big things like moving to Costa Rica to start a surfing business (save these for when we talk about vocation). Instead, look for something you can do without a big bank loan, like exercising more or spending more time with your partner. Put an asterisk by the two or three most manageable things. Stop there for now.

Take the Next Opportunity to Move One Small Step Toward Your Good Life

One day, Jonathan the guitar wannabe was on a work trip and had a long layover in Atlanta, where his friend Jocelyn lived. She picked him up at the airport, and because it was a beautiful day, they went to hang out in the park under a tree. For the hell of it, Jocelyn brought her ukulele along. She played some songs, including one she had written herself.

Then Jocelyn handed Jonathan the uke. She taught him four chords in about five minutes. There, under that tree, in that park, on a sunny day, Jonathan was suddenly strumming and singing with a friend.

On the way back to the airport, they stopped at a music store and Jonathan bought his own ukulele for eighty dollars. When he got home, he played around with his uke while watching TV. Before long, he was learning songs from YouTube. After a couple of months, he went to a Meetup group of fellow ukulele players. Later, he found a teacher.

Jonathan still had his busy life. But rather than trying to force into his life an instrument he didn't have time for—the more complicated guitar—Jonathan stumbled on an easy instrument that actually fit in the small amount of time he had available. Now Jonathan counts himself as a musician and his so-called pipe dream has come true.

If you walk around Brooklyn's Prospect Park on a sunny weekend day, you'll see that the ukulele approach to making music fit into a busy life has worked for many others, too. People say it is because the ukulele has become trendy, but it is more because the trend has allowed many people to see that ukulele is an easy entrée into the music world—a manageable first step.

Finding easy entrées into living according to our values—regardless of time constraints—is what I call the *ukulele approach.* Let's stop blocking ourselves because we don't have time, metaphorically speaking, to learn guitar. Instead, to begin living according to our values, to make music in all areas of life, to begin on our lifequest, let's look for the easy start. That's the ukulele approach.

Identifying Easy Parts #2:

Take out that list of improvements you wished you could make to your life and look at the two or three items you put an asterisk by. What is the one small action that would take you toward an improvement on your list that you could take today—the equivalent of picking up the ukulele instead of the guitar? Now do it. Then look for another ukulele to pick up tomorrow. This is how to identify the easy parts. It's that simple.

Taking the Ukulele Approach to Our Relationships with the World

Remember how, in Part Two: Wanting What You Really Want, we learned that the very act of wanting certain goals in your life—like community feeling and health—could make us happier than the same act of wanting more materialistic things? The best life relationships help us move toward (or at least don't frustrate) the fulfillment of more than one of those intrinsic goals at one time.

The fact that some transactions actually frustrate our intrinsic goals helps us understand why we feel bad when, for example, we buy certain products. If we purchase something whose manufacture we know harmed someone even far away, then purchasing that product actually stands in the way of community feeling. We feel less happy. On the other hand, if we find a way to buy a product we like and know that the purchase helps someone, we work toward affiliation. We feel happier.

The idea behind starting with the easy parts or taking the ukulele approach to the lifequest is that as you slowly bring more and more relationships in line with your values, more and more of your life itself lines up with your values. Meeting our intrinsic goals in our relationships—one at a time—is the gentle way of moving toward the Good Life.

Some Ways We Relate to the World

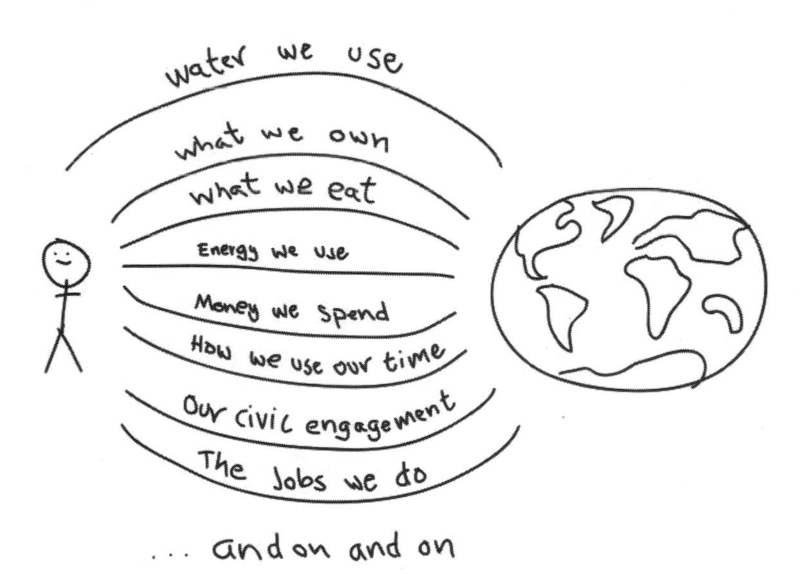

Take a look at the diagram of a typical person's relationships to the world. (Or you can draw one for your own life, if you haven't already. See directions on page 65.) Through the tiniest of our world relationships, we affect both our own happiness and the happiness of the world. Our iPhones connect us to children working in factories in China, our food choices to domestic animals everywhere, and our oil use to Iraqi and Afghani widows and widowers.

For so many of us, these relationships are out of line with our values and so we feel bad when we look at them. In our mistaken feelings of powerlessness, some of us develop a habit of suppressing or ignoring thoughts and feelings about these relationships. We stop listening to the little voice inside ourselves and put ourselves to sleep. We become like Jonathan, who ignores his desire to play the guitar because he hasn't yet found the ukulele.

The problem is, you can't selectively put yourself to sleep. You can't be asleep to the tiny voice when it nags you about the world but awake to it when it tries to tell you what direction would make you happy. When we stop listening to our inner voice, we put ourselves to sleep and we alienate ourselves from our Selves.

Part of what motivates the lifequester is wanting to put an end to the numbness. Wanting to not turn the Self off but tune in to it. Wanting to be alive. Because as we learned in the discussions in Part Two, self-alienation is the opposite of what both makes us happy and makes us able to help the world.

This isn't to say that the burgeoning lifequester looks over this diagram of his or her relationships to the world and becomes overwhelmed with grief and sadness. Being alive to your Self does not mean you have to constantly bathe in the sorrows of the world.

But instead of turning the sometimes painful feelings off, we have to come to terms with them. To be able to feel compassion for the suffering in the world without either feeling overwhelmed or having to go numb. To be able to be in the presence of our compassion. One way to be able to be in the presence of that sorrow and compassion is to be able to reassure yourself that you are taking some action.

Which is where the ukulele approach to our world relationships comes in.

The first thing to do in looking at this diagram is to see the opportunity in it. We can make small adjustments in our relationships to everything from our stuff to our food to how we travel and actually have a positive impact on the world and the happiness in our own lives. We can transform some of our relationships from chores whose consequences we prefer to ignore to opportunities to make our marks, to have meaning in our everyday routines.

The trick is to find your ukulele, metaphorically speaking. Find relationships you can fix without too big a time cost. Or even some that win you time.

For example, if you take the elevator all day long as you move between floors at work and you go to the gym to use the StairMaster™, you may win time back if you take the stairs at work instead of waiting for the elevator. Then you don't have to go to the gym and you save the world some energy. Making small changes like that—in ways that are both good for you and good for the world—is how the lifequester applies the ukulele approach.

What is some small area of your life that is not in line with your values that you have been meaning to change? Since the essence of this work is about returning to your own True Self, you are not looking for relationships that *someone else* thinks you should change, but ones that matter to you.

Identifying Easy Parts #3:

A lot of us already have some potential change that has been nagging us, but we haven't taken the action because we are frightened it won't have any real value to the world. That it is pointless. In such cases, we have overlooked the fact that—whether we help the world or not—we, at least, get a happiness dividend just from acting in line with our values. Just from being awake to our Selves.

Jot down a few of the things in your life that hurt the world in ways that nag at you.

Choose the easiest one.

Change it just for today.

When you get that item right, repeat with the next one.

Nineteen Easy Ukulele Actions You Could Take Today

Choose one thing from the list below and do it *just for today.* Then try another thing tomorrow. Make your own list. Choose actions that are fun or do you some good, too, or deal with problems that have been niggling at you. If what you do today makes you feel good, repeat it tomorrow. Remember, it is not about changing the world but about finding an easy start to making your own relationship to the world a happier one.

1. Offer to carry an older adult's groceries.
2. Ask a single parent if she needs babysitting help.
3. Handwrite—so it doesn't look copied and pasted—a letter to your congressperson asking him to do something about climate change.
4. Walk somewhere instead of driving.
5. Take some food to the neighborhood soup kitchen.
6. Collect and recycle twenty aluminum cans.
7. Unplug everything not being used at home and the office.
8. Pick up litter as you walk down the street.
9. Give at least a quarter to anyone who asks you for money.
10. Turn down your heat or air conditioning.
11. Call up the local animal shelter and ask what help they need.

12. Thank a veteran, teacher, mail handler, bus driver, or public servant.
13. Take some used books to the library.
14. Compliment everyone you talk to.
15. Eat no meat.
16. Buy a piece of pizza for a homeless person.
17. Avoid bottled water and drink tap water (filter it if you want to).
18. Call your governor's office and ask her to ban elephant ivory sales in your state.
19. Ask your employer what the company is doing to combat institutional racism.

How the Ukulele Approach Helps the Lifequester

Let's take the case of a lifequester named Annie, because Annie's experience epitomizes the process of so many lifequesters I have met. Annie lived in Brooklyn, New York, and she often felt that she could spend a tiny bit more money to buy ethically produced coffee but always shrugged it off because she felt it was too tiny a drop in an infinitely large bucket.

Annie had read that coffee is a huge export from the developing world, second only to petroleum. She'd heard that more than 25 million laborers and farmers globally work to produce coffee and there are big issues surrounding its production, including perpetuating cycles of poverty, child labor, clear-cutting of forests, and others. Inside herself, she had often felt her relationship with the coffee farmers and children and forests was not as she would like it to be.

For that reason, Annie decided to take the ukulele approach to her relationship with coffee. She decided to choose one small action she could take to improve that relationship and to do it today. The ukulele approach means not waiting till you can play guitar, but just doing what you can to move a small distance in the direction of a better life for yourself and for the world. Today.

Annie decided her ukulele action would be that when she next went to the store, she would look at the labeling on all the coffees for sale and choose the one that looked most in line with her values. She wouldn't make a mountain out of a molehill. She wouldn't spend hours on research. She would just do her best in the moment.

When she got to the store, she found shelves full of different brands of coffee. She looked at a range of the popular store-bought coffees like Maxwell House and Folgers and couldn't find a reference to fair trade or sustainability anywhere. She picked up a couple of more expensive brands and saw that they were labeled as USDA Organic or Fair Trade Certified.

Since this was her first day of changing this relationship, Annie did not know what the different certifications meant when it came to coffee. So what? She was doing the best she could in the moment. Jonathan didn't research ukuleles. He didn't figure anything out. He just went to the store and bought the ukulele that was there. So Annie just decided to buy the best coffee she could.

Standing there in the coffee aisle, she saw that some brands had one sort of ethical certification and others had many. She decided the more certifications, the better. She happened to find a coffee that was roasted in Brooklyn and was certified organic by the USDA, fair trade by Fair Trade USA, sustainable by the Rainforest Alliance, and kosher by Kosher Supervision of America. That's the one she got.

She knew she hadn't changed the world, yet something shifted within her. This was the first time in a long time that she had bought coffee without having to ignore a micro-twinge of guilt from a little voice inside her. This time, she didn't have to close anything off. Maybe she hadn't gone all the way, but the point is that instead of turning off the niggling feeling, she listened to it and acted on it. She took one step toward being awake and True to her Self and happy and one step toward helping the world.

That's the ukulele approach to the lifequest.

But how is buying a different brand of coffee going to change your or anybody else's life? you may ask. Does buying a different coffee really do that much for Annie besides allow her to give herself a pat on the back one day? Yes! Because this is only the first step in an ongoing process.

For Jonathan, buying a ukulele and playing it at home by himself began such a process. As his knowledge, talent, and passion grew, ukulele playing automatically began to push his life's time-suckers out of the way—like watching TV and clicking aimlessly around the computer. By giving a small amount of space to something he was passionate about, Jonathan found that his life slowly adjusted to make more room for his passion as it grew. Next thing he knew, he was out making music with new friends.

A similar process occurs when you take the ukulele approach to the lifequest. Take the case of our lifequester Annie. After trying her new brand of coffee, a friend told her he'd heard certification schemes sometimes have problems. Annie did some research and learned that certification costs too much for some small coffee growers who might care for their workers and their land more than the owners of large "certified" plantations do.

Annie read that some coffeehouses actually develop direct purchasing relationships with small coffee farmers. Known as "direct trade" coffee, this approach forgoes the certification process because the coffeehouse owners themselves inspect the farms for ethical and environmental practices while cutting out the middleman and paying a fair price directly to the farmer.

Annie searched for such a coffee place in Brooklyn, found one, and decided to pay a visit. She had a conversation with the café owner and found that his purchasing policies were in line with her values. She made it her habit to buy her coffee for home there and generally hung out there a bit when she went.

She sat at the counter and, over a series of visits, got to talking to and became friendly with a couple of the baristas. Her new friends were particularly passionate about social responsibility (that is why they worked there). Annie ended up tagging along with one of them when she volunteered at the local community garden. Later, Annie realized her new friends were vegetarian, an idea she had toyed with but didn't know how to pursue in her life. Her new friends encouraged her to try, and she did.

Without even realizing it, Annie had progressively applied the ukulele approach to more and more relationships in her life, a process I have witnessed happen to literally thousands of people.

When this process occurs, lifequesters enter a kind of choice-making school—experimenting with different ways of living. Putting a toe in different kinds of pools. Breaking away from the standard life approaches and following their passions instead. Meanwhile, they learn that they have agency and power over their own lives and worlds. They meet people who share their values.

Why would this make you happy? Because you are making choices according to who you really are and letting your True Self emerge. Some choices will be ethical—like Annie's coffee choice. Others will be fun—like Jonathan's choice to actually play ukulele. You are both getting from the world and giving to the world in your relationships. Your intrinsic goals are getting met. Barakah—the blessing of both sides of a relationship—is present.

By this simple process, you apply the most fundamental principle of this book—give more energy to what is true for you; give less energy to what isn't—and begin to build a life based on who you really are.

How the Ukulele Approach Helps the World

Old-fashioned climate activists used to say that wearing sweaters and turning down the heat in order to use less fossil fuel would not save the world. Instead, they wanted us all out marching for the government regulation on climate that would cause widespread societal change.

But my friend Juliet Schor, the sociologist and author of *Plenitude: The New Economics of True Wealth* and many other books, writes with her colleague Margaret Willis in a research paper published in the *Annals of the American Academy of Political and Social Science* that what begins as "political consumption" for a person very often ends up being just one part of a "larger repertoire of strategies and actions oriented toward social change." Willis and Schor's research shows that consciously making small life choices in line with a desire to help the world is highly correlated to higher levels of activism and civic engagement.

In other words, people who begin by wearing sweaters and turning down the heat ultimately turn out to be more likely to go on climate marches. In fact, if you are interested in social change, it is worth considering the fact that convincing people to wear sweaters and turn down the heat may be a way to ultimately get them marching. Becoming more responsible in our personal lives takes us through a process that makes us more responsible in our community and civic lives.

I'm not saying that any of us, as lifequesters, must necessarily become activists. I am saying that committing to what we have called "changing from the outside in" or the "easy parts" or the "ukulele approach" is neither trivial nor self-centered. It is the beginning of a self-perpetuating path that leads to a deeper and more compassionate and engaged relationship to the world.

People who change one of their world relationships end up on a path that leads them to change more of their world relationships and to influence others to do the same. That, when amalgamated with similar efforts by other people, amounts to the beginnings of cultural change—shifting of the norms by which all of society lives. In other words, we lifequesters aren't just drops in the ocean but drops who, together, actually change the ocean.

How does cultural change contribute to the mechanics of robust societal change on the scale of, say, emancipation, suffrage, and civil rights?

The truth is that, while there are many strong opinions on the subject, no research yet exists that can tell us exactly how society changes. As a result, none of us gets to know for sure which of our actions will have the greatest effect on the world. Rosa Parks certainly did not know that she would be remembered forever when she simply chose not to give up her seat on the bus.

What we do know is that, in history, societal change seems to come in a sort of transformative two-step. First, there is cultural change—people become less homophobic, say. This leads some U.S. states to recognize "domestic partnerships"—legislative change. Legislative change has the knock-on effect of making homophobia even less socially acceptable—more cultural change. This leads politicians to the first state recognitions of gay marriage—legislative change.

The mood of the public shifts, and then the legislators do something to support that shift, which in turn alters the mood of the public further, which in turn leads to more legislative change. To ask which of the steps in this dance—the cultural or the legislative—is more important is a little like asking whether the right leg or the left leg is more important when walking.

Meanwhile, what makes for the cultural change that is part of the two-step is the amalgamation of lots of people changing their behavior and lives. That includes, of course, lifequesters taking the ukulele approach.

When you start your quest, you are just starting to, say, buy coffee differently. You are actually at the beginning of changing the relationships in your life so that they are in line with your values and so your actions and routines make for happiness for you and happiness for the world you live in. The ukulele approach is like the low end of the ramp that leads people not to one but to a whole group of strategies for fixing their lives and the world.

If you think about it, this is an incredibly hopeful message. You don't have to be a martyr. All you have to do is what you can do today. That takes you to a new tomorrow when all you have to do is what you can do that day. Strain is not required.

Ultimately what we are after is not an imposed prescription for either individual action or collective action but the ability to respond and be *alive* to ourselves, our compassion, and the best route to bringing happiness to ourselves and the world, based on our own individual capacities.

Six Lives Transformed by Taking the Ukulele Approach

Sarah Stopped Using Plastic Bags

Bisbee, Arizona: *I wanted to take a small action in line with my environmental values, so I started to bring my own bags to the grocery store. I began making reusable bags out of recycled materials and then I began advocating for a plastic bag ban as a community organizer with Bags for the People. Three years of hard work later, Bisbee, Arizona, banned plastic bags and people always look to me as the girl who took one small step for herself and created good change.*

Persille Planted Food Scraps on Her Windowsill

Copenhagen, Denmark: *I started by growing the small, root part of a spring onion by just putting it in a pot with soil. My record is eating the same spring onion eleven times. I also regrow cabbage, lettuce, carrots, celery, fennel, turmeric, ginger, beets, leeks, and onions. And even scraps like seeds from lemons and oranges and the stones from avocados and dates are growing in my apartment, though they can't produce fruit in my neck of the woods. Now I teach others to do the same thing through my blog and Facebook page, I've made connections across Scandinavia, and I'm writing a book.*

Derrick Volunteered as a Firefighter

Seaside, Oregon: *When we moved to town, I thought volunteering would be a great opportunity to meet new people. After a while, a couple of friends in the fire department suggested I try out for the emergency medical team. Then I went back to school and got my certifications and became a full-time emergency medical technician, riding ambulances. It was a great blessing to achieve a sense of community and involvement.*

Lorraine Rescued an Injured Pigeon

New York, New York: *Twice in eight months, my upstairs neighbor found injured birds and brought them to me because I have a pet bird. Both times, I brought the birds to the Wild Bird Fund. They treated them and later gave them back to me to release. It was just a wonderful feeling letting those birds go free. I wanted to help more wild birds recover and find their way back to their lives, so I volunteered mostly cleaning cages and feeding birds, and then I got my state-issued wildlife rehabilitator license. Now I spend my days staring into a microscope to help figure out why the various New York City birds might be sick so we can get them back up and running.*

Troy Switched to Free-Range Eggs

New Brunswick, Canada: *I became a vegetarian but still ate eggs for protein. By age thirty-five, I had already had two cancer scares. I didn't trust the food system. I started looking for local sources for my free-range eggs and ended up buying ten hens. We haven't bought eggs since and usually give a dozen away each week, too.*

Kim Went to a Community Meeting

New York, New York: *I am originally from Pennsylvania and I had heard of drinking water being poisoned by fracking for natural gas there. I heard that the gas industry was building pipelines to bring fracked gas into New York and I went to a community meeting about it. I ended up becoming heavily involved in Occupy the Pipeline, left my job in fashion, and now work as a coordinator with the Sane Energy Project.*

An Easy Change of Focus—Not Just for Me

Remember my friend Julia Butterfly Hill, best known for her activism against logging by living in the branches of a 1,500-year-old giant redwood tree? She has a saying: "It is not that we have to fundamentally change our Selves. We just have to change our focus."

Working on the "easy parts" means we don't necessarily need to find time to build new relationships with the world or cause some sort of upheaval in our lives. It means we simply change the focus in our existing relationships to ask how we can maintain them in ways that are better for us *and* better for others—not just for me.

In this way, we make the things we have to do every day more meaningful for ourselves and find that we have choice in things we might otherwise feel obligated to do.

Sometimes, though, because we are so used to doing things a certain way, it is hard to know how we might change our relationships to make them more authentic. We are not sure where our metaphorical ukuleles lie. That's why, in the next few chapters, we are going to look at some of the relationships common to all of us—with food, with stuff, with transportation and housing, with our civic structures—and see how other lifequesters have adjusted them to make them better.

In that way, we can learn to feel free and powerful and authentic and True—not by totally changing our lives but by adjusting the focus in the lives we already have.

13

EAT HOW YOU WANT TO LIVE

Portrait of the Good Life at the Dinner Table

Late on a Saturday afternoon, you stroll with a couple of friends through a farmers' market. It's fall, say, so while you chat you pick up a good-size bag of root vegetables—onions, carrots, potatoes, beets, turnips. The three of you go home to your house and chop and talk and toss the veggies in olive oil and salt and pepper and throw them in the oven on a baking tray (450°F, in case you want to try).

When the veggies turn brown and a little blackness tinges some corners, you take them out, throw them in a pot, and carry them over to another friend's house. One or two other little pods of friends have congregated. Some of them, too, spent a couple of hours chopping or stirring or baking or sautéing together. You all chat and eat and maybe someone breaks out a guitar and you sing.

You're having a feast.

After a while, someone says they crave a sweet and you all stroll to the ice cream store for a single scoop on a cone (in other words, not a whole pint). Maybe you stop for a drink and then meander back to your friend's house to collect your pot and your leftover veggies. People give you a little of what they cooked to take home, too.

That's Saturday—kind of a routine. Maybe a friend started it on Facebook or by e-mailing out a few invites and it turned into a regular thing.

On Monday, you throw some of the leftovers from the party in a great big mason jar and take them to work. At lunchtime, you invite a colleague to join you, find a table, and sit and chat and eat. The next day, inspired by your kindness, your colleague brings you lunch and you eat together again.

In the evening, you throw what is left of the veggies in a blender with some water and soy sauce, you heat it up and you've made a lovely soup that, with some cheese and bread, makes dinner for the next couple of days.

Not until Thursday or so do you begin to run low on the twenty-five dollars' worth of fresh food you bought last Saturday. Because you spent so little, you can splurge on a good local-food restaurant on Friday night. On Saturday, you venture back to the farmers' market, wondering what you'll find this week.

Sound like a fantasy? It's not. I know so many lifequesters who—by taking tiny ukulele step after tiny ukulele step—finally found themselves relating to food this way. Mealtimes are not a struggle but a joy. Patiently making tiny changes, one step at a time, they've found that exactly what they want from their Good Lives is reflected in how they eat.

Then There Is the Average American's Approach to Food

Let's say you got too busy with all the work e-mails and stuff crowding out your life and totally forgot to make plans for Saturday night. Maybe you don't mind, because you are exhausted anyway and tired of trying to do the right things at the right places with the right people.

Besides, assuming you're anything like what research says about the average American, you prefer to eat at home. But also—like an average American—you rarely cook a meal from scratch. You buy takeout or prepared food from the grocery store's frozen section. Tonight, like one in eight Americans, you plan on pizza. If you are not going to be having a great time out with friends, you at least deserve some comfort food, right?

Like most Americans, you sit and eat in front of the TV (one-third of Americans say their TV is *always* on during dinner). Chances are, to stave off the loneliness of eating alone, you are e-mailing or texting from your smartphone, too. You suddenly realize you've finished your pizza, but you don't even remember eating it.

When the ads come on, you run to the freezer and pull out a pint of Ben & Jerry's. Americans eat forty-eight pints of ice cream per person a year—more than any other country—and two-thirds of Americans most frequently eat their ice cream in front of the TV or on the couch. Hell, the ice cream is gone and you don't remember eating that, either.

Sunday morning comes and there are no leftovers to eat or share during the week. What is left, instead, is a pile of cardboard and plastic trash, a self-imposed guilt trip, and an obligatory visit to the gym to try to burn off the calories by running in place on the treadmill. It is not as bad as it could be. At least this meal was not eaten in the car—like one in five American meals are.

What Comes on the Side with That American Meal?

If you're like me and you had the choice between the feast with friends or the frozen pizza dinner options, you'd choose the feast every time. You get to eat food that is actually good for you, have fun, take care of your body, share with and help other people, and feel like you are both giving to and getting from life. You'll read below that this option is also way better for food workers, animals, the land, the oceans, and the climate.

This kind of eating epitomizes the kind of two-way world relationships the lifequester is looking for—full of security and meaning, bursting with barakah, and meeting our intrinsic goals. Trust me for a second when I say that developing a relationship with eating that is in line with a happy, meaningful life is way easier than you think.

On the other hand, ask yourself if you like and enjoy what comes on the side with the standard American diet:

Disease: One in four Americans eats some sort of fast food every day. Our favorite food is hamburgers. We eat 31 percent more packaged food than fresh food. Meanwhile, 60 percent of Americans weigh more than is considered healthy. Four of the top ten causes of death—coronary heart disease, diabetes, stroke, and cancer—are related to how we eat. *How much healthier would we be if we all ate happily and healthily?*

Poverty: Pizza makers and burger flippers—food-preparation industry workers—account for 47 percent of those who work for minimum wage or below in the United States. As a result, more than half of fast-food-industry workers rely on some form of public assistance. A food system based on creating food that was actually good for us, on the other hand, would likely provide more and better employment. *How much less would we contribute to poverty if we all ate happily and healthily?*

Environment: Thanks to the agricultural chemicals washed off the corn-belt farmland and into the Mississippi River, there is a five-thousand-square-mile dead zone in the Gulf of Mexico where no fish can live. Think of what those chemicals do around the world. Meanwhile, animal agriculture alone is responsible for 18 percent of greenhouse gas emissions. *How much safer would our planetary habitat be if we all ate happily and healthily?*

Loneliness: Americans eat more than half their meals alone, and 60 percent of us eat breakfast alone. Forty percent of American adults say they are lonely. *How much less lonely would we be if we all ate happily and healthily?*

Do It Now!
Easy Food Step #1

The consciousness brought to eating by planning and getting ready for a group dinner is arguably the entire basis for a happy and healthy relationship with food.

My Facebook follower Cynthia Gubbish from Bronx, New York, wrote to me: "The best thing about eating is cooking together, the company, sharing the creative process, and all the conversational chopping. When it turns out tasty, that's the cherry on top."

Right now, e-mail five to ten friends and suggest that you all grocery shop, cook, and eat together next Friday or Saturday night.

Don't worry about how healthy it all is. Just follow this one rule: no prepared food allowed.

Stop reading this book for a minute. Send the e-mails. Do it now!

Who Started This Food Fight?

The standard approach to eating puts us in a fight with food that we don't have to be in. The marketing tells us there are magic foods that will make us super healthy or that will make us feel really good or that will be such a pleasure to eat. The claims that the advertisers make contradict each other, change from season to season, and leave us confused. As a result, we wrestle with how much our food costs, what is good for us, what tastes good, how long it takes to prepare, and what its production does to the world.

Standard Eating Approach A—Power Food: Sometimes we approach the fight by imagining our bodies to be mechanical machines that need the highest-quality food for fuel. We don't enjoy eating it any more than a car enjoys going to the gas pump.

We just try to get the highest octane—like from expensive fresh-squeezed vegetable juices, smoothies, and supplements.

Standard Eating Approach B—Orgasm Food: Sometimes we worry less about nutrients and instead try to wrest from food some sensual pleasure—beautiful food on beautiful plates in beautiful places surrounded by beautiful people. Indeed, writers and commentators talk about "food porn" and some say our obsession with the orgasmic food experience may come from the fact that too many of us don't get enough sex—or at least not enough of the right kind.

Standard Eating Approach C—Comfort Food: Other times we try to get another thing that food itself cannot give—comfort. The problem is that neither a potato chip nor a chocolate bar can actually comfort us. Comfort at a meal comes when someone you love sits and eats with you and talks to you and listens to you and cares for you. Comfort cannot really come from a pile of sugar-laced frozen fat on top of a waffle cone.

Do It Now!
Easy Food Step #2

You know how sometimes you eat a carrot and it's so good you're like DAAAAMN...

—BESS ROGERS, GUITAR PLAYER, THE SECRET SOMEONES

Never mind about your diet. In fact, never mind about your relationship with food. All you have to think about is your next meal or snack. Go out to a store. Get an apple or two. Or some carrots or a cucumber or a banana. Whatever it is, decide that your next snack or meal will include something that did not come in a package.

Do it now!

So the food fight goes like this: We munch a food-like product that the ads say will fuel our bodies. It isn't very satisfying and we feel deprived. To cure that, we chow down on what the ads say will bring pleasure. That makes us feel guilty, so we then snack on foods they say will comfort us.

Each food product makes up for the shortcomings of the other food product, according to the food industry's $32 billion marketing budget. We eat one food product to cure us from what the last one did to us. Meanwhile, we don't get cured at all. We become overweight and ill.

According to a 2012 study by the International Food Information Council, 52 percent of Americans believed doing their taxes was easier than figuring out how to eat healthily. Some people feel so confused about their relationship with food that they no longer like to eat in front of others for fear they will be judged. That is the food fight we are in.

Here is how the food industry explains it:

> **Baloney Explanation #1:** *"Our diet is a casualty of our wealth and choice. We are victims of our own success as a species."* Who wouldn't get fat when faced with the incredible abundance of forty-four thousand items sold in the average grocery store? We should be grateful that the world is like one big Willy Wonka chocolate factory.
>
> **Baloney Explanation #2:** *"We have evolved to gorge and store calories as fat to prepare for famine. But famine no longer comes."* The problem is that some of us are too weak-willed to fight our own evolutionary biology. We should accept that there will always be a fight with food because we don't live in the jungle that our bodies are designed to live in.
>
> **Baloney Explanation #3:** *"We only give customers what they want. If we didn't produce it, someone else would."* Wait, um, if we want it so bad, how come your advertising budget is

bigger than the entire GDPs of more than half the countries in the world?

The thing is, if these explanations were true, why wouldn't the citizens of other affluent countries like Japan, France, and Italy be in a similar fight with food? They have just as much choice and just as much food available. Why isn't their food also making them sick and fat and wrecking the planet?

Here is what the food industry might say: All those countries have traditional diets that evolved over thousands of years. The United States is such a young country filled with so many ethnicities and national backgrounds. The problem with our diet is that we have been so successful at making a country that includes everyone, and a hodgepodge diet is a small price to pay.

But this argument is baloney, too.

It is true that each of the other countries mentioned has its own mix of foods that seems, at first glance, completely contrary to the others. The Japanese diet relies heavily on fish, while the Italian relies heavily on pasta. How can these contrary approaches both end up healthy? There must be some secret combination of nutrients that can't be replicated without a passport and an airline ticket.

It all just seems too mystical and complicated, right?

Wrong.

As Michael Pollan so excellently points out in his book *In Defense of Food,* what all these diets have in common is that they all consist of real, unpackaged, unprocessed, unmanufactured, unchemicalized food. This is why the first sentence of the book simply reads, "Eat food"—as opposed to food-like products. What makes other national diets healthy is not so much the traditional combinations of foods but that the food itself is "traditional"—meaning that it comes from a farm, not a factory.

If you think of a traditional diet as one that embraces traditional, farm-produced food—as opposed to food-like products—then it has been only forty years since the United States had a traditional diet, too. When we did eat real food, our diet was much healthier. It was when our admittedly hodgepodge diet moved away from real food to food-like products that our fight with food began. Only then did our diet begin to make us and our habitat sick.

Nor was it something in our character that made us take a wrong turn in our diet. It was a deliberate and aggressive change in agricultural policy that began back in 1971, when President Nixon appointed the agribusiness crony Earl Butz to head the United States Department of Agriculture (USDA).

Butz vigorously pursued the formation of an industrialized, centralized, hyperefficient food system in which the world would be fed on different concoctions made from corn and soy, crops easily adapted to factory farming. To small family farmers who had the tradition of putting stewardship of land, food, and animals first, he insisted the time had come to "get big or get out."

The availability of incredibly cheap soy and corn resulting from Butz's policies also meant livestock could be fed on cheap grain instead of grazed on grasslands. This led to an explosion of concentrated animal feedlot operations (CAFOs), which meant lots of extremely cheaply—if very cruelly—produced beef, pork, and chicken.

As a result, according to Michael Pollan, while the cost of fresh fruit and vegetables has risen 40 percent since 1980, the costs of the standard ingredients of packaged snacks, soft drinks, and prepared microwavable entrées has gone *down* 20 percent. Suddenly, Americans began buying boxes full of food-like substances and adding an extra three hundred calories to their waistlines daily.

Meanwhile, since the only thing this fake, food-like substance really has going for it is so-called convenience—meaning you don't have to cook—we began eating alone and on the go. We gave up on a meal consisting not just of eating but of sitting and talking and being together.

The good news is that there are burgeoning movements of food enthusiasts who embrace all sorts of healthy and kind eating practices—from the local, slow food, and organic movements to the vegetarian and vegan. These movements are filled with lifequesters who want to get their relationship to food and animals and the world right.

They have rediscovered that what makes a meal great—what fuels our hearts and souls—does not come from the food alone. They are returning to what the Japanese, French, and Italians never left: meals where you cook real food and then sit with friends and family to eat it.

And through their happy, healthy relationships with eating, they show us that, while we didn't start the food fight in our lives, we can absolutely end it.

What to Eat in Twenty-Three Words and a Few Footnotes

Healthy, happy, authentic eating is a lot about *how* we eat, but let's start with *what* we should eat. It is hard to believe how simple it is. You don't need to read a book or go to a gym or see a nutritionist or order any vitamins.

It is so simple, in fact, that I feel compelled to tell you that I did not make it up. Instead, it is distilled from the writings of the two people I take food advice from, the respected food writer Michael Pollan and the respected nutrition writer Marion Nestle:

Eat real food.[1]
Not too much of it.[2]
Make it mostly plants.[3]
Go easy on junk food.[4]
Forget about nutrition.[5]
Use your body.[6]

1. If it has ingredients a farmer didn't grow, it's not real food.
2. Eating too much is pretty hard to do if you snack on cucumbers and carrots.
3. The "need" for meat and dairy is a food industry con (see below).

4. If there is an advertisement for it anywhere in the world, it is likely junk.
5. Eat different colors of food—red, green, orange—and you're fine.
6. Walk, run, ride your bike, or take the stairs so you are puffing at least a little for twenty minutes every day.

Eat What You *Really* Crave—Advice from the Costar of *Super Size Me*

You may remember Alexandra Jamieson as Morgan Spurlock's costar in the documentary *Super Size Me,* which followed Morgan as he spent a month eating only McDonald's food to see what would happen. Alex is the author of many food books and, most recently, of *Women, Food, and Desire: Embrace Your Cravings, Make Peace with Food, Reclaim Your Body*. She is a chef and "functional nutrition coach."

Alex says we should not choose some food-plan dogma to adhere to—another standard life approach—but learn to listen to our authentic selves and to eat what we *really* crave:

The first thing to say is that eating food is one of the best things about being a human being. We have these incredible bodies that allow us to taste and touch and smell but we forget to enjoy. We forget to enjoy our own senses.

We live so much in our heads that our bodies feel deprived. We don't know how to taste the simple foods or be happy without spending lots of money. We've forgotten how to savor the taste of a simple apple.

So when I say eat what you crave, it is not just about choosing the food you want to grab for right this second. First, you need to know what truly makes you happy. It is generally not the striving and the acquiring or gorging.

It is not about denying your desire but about sitting quietly and listening to your body long enough to let it tell you what you actually want.

More than some kind of junk food, most people crave more energy, focus, strength, confidence, and peace. When you get in touch with those deeper and broader cravings you can ask, If life energy or peace is my big motivating desire, what food do I need to eat to have that? What could I eat now that would help me move in that direction?

Most often it is not the pie or potato chips. Mostly, it is real food that helps you stay in the reality of what makes for happiness. You know in your body that eating junk food isn't going to help you stay in a place of peace (though sometimes it is what you want just for a pleasure boost and that's okay, too).

When it comes to eating badly, the trigger may be stress or overload or even car traffic. Watch for those things that trigger you and transform the action into something that will actually help you. What can help you actually relax rather than just anesthetizing yourself with a cookie or the drive-through?

Maybe it's as simple as getting up and going for a short walk instead of forcing yourself to sit still at your desk. If you have a meeting with a colleague, you can even suggest to her that walking together might make your brains work better than sitting at a conference table together.

We aren't used to really listening to ourselves, so it takes time to learn this, to know what you really want in your life and so what you really want to eat. So watch for the triggers and transform your responses, one at a time. Be patient with yourself. Before long, you'll learn to eat what you really crave.

How to Eat the Way You Want to Live

Here are two questions to ask yourself:

1. What would my food life look like if it were in line with my values for myself and for the world?
2. Given my life circumstances, what is the first tiny step (the ukulele, if you like) I can take today to get there?

A good relationship with food nourishes not just our bodies but our hearts and souls. At the same time, choosing our eating habits well also offers a path toward living more in line with our values and away from the terrible human exploitation, animal suffering, and planetary destruction of Big Food.

You may think that the portrait of the Good Life at the dinner table that we painted sounds impossible given the circumstances of your life. But why should having a healthy, fulfilling, fun, service-oriented life sound like fantasy? After all, what are you actually alive for?

Try picking from the tools below. Nobody says you have to do them all at once or that you have to do any of them all the time. But practicing one or several may move you toward the type of eating you want. Don't turn it into a grind, since the grind is exactly what we are trying to get away from. Remember, we are into the gentle approach—stress free and easy.

IF YOU'D LIKE TO BE MORE PRESENT: EAT LIKE YOU MEAN TO

When you take a handful or a forkful or a sip of anything, stop. Turn off the TV. Put your phone and computer far away. Stop walking. Sit down. Experience. Enjoy. Be grateful. And if you are paying attention and you suddenly realize you are not enjoying what you're eating or you don't really like what's in it or how it was produced, you know what you will do? Either stop or eat something better.

IF YOU WANT TO ENJOY MORE FRIENDSHIP: NEVER EAT ALONE

As a practice, if you get a cupcake, cut it in half and share with a friend. Pull your chair up to a colleague's desk at lunchtime. If you drink a coffee, sit down and sip it with a friend. Eat meals together with your roommates or family. Ask some friends over for potluck dinner. Mealtimes are not just about food but about resting and joking and socializing and talking and sharing. Research shows that people who eat together automatically eat better.

IF YOU FEEL CRUNCHED FOR TIME: COOK IN A BIG POT (FROM SCRATCH)

A 2003 Harvard study shows that most obesity in the United States can be explained by the rise of eating prepared, so-called convenience foods. But how can they be convenient if they make us fat? For real convenience, cook from scratch for yourself once a week in a big pot and then eat leftovers. This will save you time scouting the frozen food aisle. You'll also eat less and choose ingredients that are better for you and for the world.

IF YOU WANT TO SHARE YOUR DISCOVERIES: MAKE SUNDAY AFTERNOON A COOKING PARTY

Cooking is most fun when you do it with a friend and easiest when you choose some leisurely time to do it—as opposed to when you are hungry and in a rush. Also, planning with a friend to cook will help you stick to your commitment to healthy eating. Invite a friend over on a Sunday afternoon and cook together for both of you.

IF YOU FEEL OVERWHELMED: LIMIT YOUR CHOICES

One reason inexperienced people don't cook is that they think it is daunting. They look at a big fat cookbook and don't have half the ingredients and don't own the right pots or pans. At first, limit your meals to things you can cook in a single saucepan made from beans, lentils, rice, pasta, bread, olive oil, garlic, onions, avocados, carrots, kale, soy sauce, salt, and pepper. It is not going to be gourmet all the time, but if you accept that you're on a learning curve you'll find you have a less finicky palette.

How to Boil Water—a First Step into Cooking

"There is a prevailing theory that we need to know much more than we do in order to feed ourselves well. It isn't true. Most of us already have water, a pot to put it in, and a way to light fire. That gives us boiling water, in which we can do more cooking than we know."

—TAMAR ADLER, *AN EVERLASTING MEAL*

Try this over and over until you do it the way you like. You'll learn how to do it right by doing it a few times. It's okay to fail. It doesn't have to be perfect:

1. Soak a couple of cups of dried beans of pretty much any sort in water overnight.

2. Peel, say, three cloves of garlic and one onion. Chop the garlic and onion up the way you imagine they should be chopped (there is no right way!).

3. Pour enough olive oil to fill the palm of one hand into a pot. Turn heat to low, where on a gas stove the flame is just above the level where you think it will go out.

4. Throw the garlic in for a couple of minutes.

5. Throw in the onions and cook until they are soft, stirring occasionally.

6. Drain the soaking water off the beans and throw them into your pot. Fill the pot with water until the beans are covered by an inch or so. If the beans swell above the water later, just add more water.

7. Turn the heat to high.

8. Throw enough soy sauce in the water to make it taste like a vegetable broth or light soup.

9. When the water boils, turn the heat down until just some little bubbles are coming to the top.

10. Let cook until the beans are soft enough to squish easily between your teeth without any resistance, probably about an hour.

11. Add some chopped carrots, potatoes, broccoli, kale, or whatever else you feel like.

12. Keep cooking until veggies are whatever softness you like.

13. Get some really nice whole wheat bread and maybe some hot sauce if you want.

14. Have salt and pepper handy.

15. Eat.

IF YOU WANT TO LEARN MORE: HANG WITH FOODIES

If you want to learn to play music, hang with musicians. To make art, hang with artists. To cook, hang with people who are passionate about food. In the United States there is probably no one more passionate about food than members of the slow food and local food movements. Both care about really enjoying food, being healthy, and producing food in ways that are good for farmers, the land, the workers, and the animals. If you care about that too, find them on the Web and hang with them.

IF YOU CARE ABOUT PRODUCTIVITY: MEASURE YOUR TIME AND MONEY

For all the convenience the prepared food industry is supposed to bring, the average American spends only about thirty minutes less in the kitchen each day than when we cooked from scratch. Meanwhile, studies show that stalking the frozen food section for pizza or venturing to the takeout place requires about the same time as it would to cook from scratch. Ask yourself how much time and money prepared food really takes.

IF YOU HAVE A DISCERNING PALATE: CHOOSE LOCALLY GROWN, ORGANIC FOOD

Conventional agriculture grows its fruits and veggies with tough skins and picks them unripe so they can be shipped. Local farmers don't ship their produce long distances—which is better for the planet—so they grow them to burst with flavor and nutrients. They also don't tend to cover their goods with chemicals that are poisonous to people and the world. To get food that is healthy and tastes the best, Marion Nestle recommends in *What to Eat* that we choose, in order: (1) organic and local; (2) organic; (3) conventional and local; (4) conventional.

A Word on Food Cost

One objection people raise to eating healthy, local, organic food is cost. However, a 2013 Harvard University study shows that the healthiest diet costs only $1.50 a day more than the unhealthiest one. This could be a burden for some families, but researchers believe the increase is likely offset by savings on the health costs and missed work that come with diet-based disease.

IF YOU CARE MOST ABOUT HEALTH: EAT FEWER ANIMALS

After former president Bill Clinton's emergency heart bypass surgery, he largely cured himself of heart disease by going on a plant- and whole-food-based diet, together with moderate exercise and stress management. Overall, vegetarians and vegans have much lower rates of heart disease and cancer. Don't worry about protein or calcium. The idea that we all need meat and dairy comes from years and years of multibillion-dollar marketing by the beef and dairy industries. We can, in fact, get all the protein we need from vegetables, grains, and pulses.

Ten Reasons Why Not Eating Animals Is Better for the World

I was nearly one hundred pounds overweight and completely out of shape. I read a book that said I didn't need to eat meat and I went for it. I ended up losing ninety pounds, became a vegan, and trained as a nutritionist and personal trainer. I work with underserved communities of Detroit to help those who need the most help. I have helped to develop nutrition programs, community gardens, and community-based workout programs to help keep people healthy. It makes me happy to help people who want to make a better life for themselves and for the planet. —KRYSTAL CASTLE, ROYAL OAK, MICHIGAN

If eating ethically is important to you, choosing to limit or avoid meat, fish, and other animal products is probably the most effective step you can take.

1. The oceans contain only one-tenth as many large fish as were there before the industrial fishing era.
2. Between 20 percent and 50 percent of the world's greenhouse gas emissions come from livestock.
3. Animal agriculture is the leading cause of species extinction, ocean dead zones, water pollution, and habitat destruction.
4. The meatpacking industry is considered to be one of the cruelest and most dangerous for workers in the United States.
5. Growing feed crops for livestock consumes 56 percent of water in the United States.
6. Livestock systems occupy 45 percent of the earth's total land.
7. For every one pound of fish eaten, an average of five pounds of unintended marine species are caught and discarded as "by-kill."
8. As many as 650,000 whales, dolphins, and seals are killed every year by fishing vessels.

9. Land cleared to grow feed and graze livestock is responsible for 91 percent of the destruction of the Amazon rain forest.

10. One day of eating only plants saves 1,100 gallons of water, 45 pounds of grain, 30 square feet of forested land, 20 pounds of CO_2 equivalent, and one animal's life.

IF YOU WANT TO UNDERSTAND YOUR PLACE IN THE WORLD: VISIT A LOCAL FARM

So many of us have never seen food growing. We think what we eat comes in boxes or cans or bags. We have only the vaguest sense of where our food comes from, and this makes the entire process of survival mystifying and unnerving. When you visit a farm where your food comes from, a sense of partnership develops. You support the farm by paying for the food. The farmers give you food that is actually good for you and the planet. Suddenly, you feel like you have a place in the world.

IF YOU WANT TO UNPLUG FROM THE MATRIX: GROW SOME FOOD

Sprouting: it's cheap, easy, and nutritional, and you can do it even if you have no access to other forms of local food. Start with dried lentils from the supermarket. Take a cup or so. Put them in a bowl. Rinse them several times. Drain them but make sure they're still moist. Leave them on the counter out of the sun. The next day, rinse them a couple of times and drain them so they are moist. Do the same each day. Don't ever let them dry out. After about three days total, they're sprouted and you can eat them by the handful.

The Importance of Doing for Yourself

Sharirashtrama: the Gandhian spiritual principle that a person who finds the time to perform the manual or "bread labor" required to meet her own bodily needs will be happier and do better by the world.

Swadeshi: the Gandhian spiritual principle that a person will be happier and do better by the world when, in cases where he cannot meet his own needs through his own labor, he relies on and trades with neighbors, rather than anonymous, faraway people and companies.

Self-reliance: the strong belief that you are capable of successfully dealing with the challenges life throws at you and that you are able to take guidance from yourself rather than from other people or things. One of the major benefits of the practice of *sharirashtrama* and *swadeshi.*

Think, for a minute, of an idealized version of a sustenance farmer who has a garden plot outside her kitchen door. Each day, she goes into the garden, harvests some vegetables, gathers some eggs from her chickens, brings the food back to the kitchen, and cooks it. Each day, she meets her own basic needs by the fruits of her own physical labor and what the earth naturally provides. This is living by the principle of *sharirashtrama,* or bread labor.

Sometimes she has extra food, which she trades with neighbors for other things she needs—say, wool from the sheep farmer and firewood from the forester. Understanding their reliance on one another, when times are hard for her neighbors, she helps. When times are hard for her, they help. This is the principle of *swadeshi,* or local trade.

She has seen how, by the fruit of a fairly small amount of labor, her garden and community provide for many of her daily needs. The connection between her work, her community, and her sustenance is simple and clear. For this reason, she feels a basic security. Life is not luxurious, but for the most part, it feels safe. The food always grows. That is the sense of self-reliance resulting from her *sharirashtrama* and *swadeshi.*

Because of the simple relationship between her efforts and her well-being, our farmer has a natural clarity, simplicity, and peacefulness in her thinking. She doesn't lie awake figuring things out because the basic facts of life are straightforward. Because she is self-reliant, she is free to be who she wants without the fear that rejection will lead to her not being safe.

Since she has a basic sense that each day provides for itself, it doesn't even occur to her that she should scheme or worry about how to get more from others. Knowing where everything she owns and uses comes from, she need not worry that someone is being harmed by how she lives. If there is more on her table than she needs in the day, she knows she can afford to be generous with strangers and visitors.

Without even trying to be "better," she experiences her own basic virtue. Without making an effort, she embodies *ahimsa,* nonviolence. Thus, living according to the commonsense principles of *doing for oneself* (*sharirashtrama*) and *relying on neighbors* (*swadeshi*), our sustenance farmer epitomizes a relationship between a person, her bodily needs, and her community that naturally makes for a happier self and a happier world.

Of course, we can't all move to the countryside and break out pitchforks. But what this story of our sustenance farmer does is help us understand why it is so challenging to feel and act peaceful in a much more complicated life where the connection between working and getting our needs met is so tangled and indirect.

Where does our food come from? How does money get in our bank accounts? What even is a bank account when there is no actual paper money involved? Without really understanding where our well-being and security come from, how can we trust their permanence? In this way of living, instead of feeling safe in the understanding of how we can always meet our own needs, some of us come to feel deeply dependent on faraway systems we don't understand at all.

To keep ourselves plugged into those systems, we push harder at the only lever within reach, the only thing we seem to have control over—our work. We get completely caught up in our jobs. We worry about whether this speech gets written or that purchase order gets fulfilled or whether we meet this other deadline.

We start to take these heady tasks as seriously as if our lives depended on them. We get so caught up that we begin to think we have no time to do the things that our lives actually do depend on—like preparing our food and taking care of our children and running our own lives. So we pay other people to do that stuff for us.

This only increases our unconscious sense of dependence. We feel we are precariously balanced at the end of a very long limb that could break at any time. We sense that we are nodes in a Matrix that might cease to supply us at any moment.

That sense of insecurity and dependence makes us forget that we live, many of us, in a luxury that even kings and queens did not have only a couple of hundred years ago. We have running water and electric light and machines to wash our clothes and flush toilets. We are clean and healthy and sleep in soft beds.

In spite of our relative luxury, we feel insecure. Because we depend on those faraway systems whose fragility we intuit, our thinking becomes complicated. We see the harm in this way of life, yet we feel stuck in it. We see people suffering, but we are too scared to be generous. We feel fear and experience scarcity and feel forced to do things we aren't proud of.

Living lives that are almost exactly the opposite of our imaginary sustenance farmer's, we struggle to feel and act with peace. We are lost from our Selves and from our fellows.

Gandhi recognized that we cannot and probably should not all live like the farmer. But in spite of the lifestyle gains of our modern world, we also lose a lot by not having such a straightforward life, by not understanding that we can do for ourselves. Partly for this reason, he believed that those who lived in the professional classes should do at least one hour of bread labor every day.

In his own case, he spent an hour each day spinning yarn for clothes from cotton. For him it was a sign of his independence from the cheap cloth Britain dumped on the Indian market. It removed him from the harm done to the Indian local economy by the imports from far away and gave him a sense of solidarity with the impoverished workers in the British cotton mills.

But we don't have to spin to regain some sense of self-reliance. We can ride our bikes or walk or build things for ourselves or do our own laundry or clean our own houses—anything that reminds us that we can meet our own challenges and from which that sense might spread.

Or we can grow and choose good food and cook it. We can remove ourselves from the corporate food system the way Gandhi removed himself from the British cotton system and get a sense of our own resilience at the same time. This is part of why it is so important to do for ourselves.

"Labor," the Russian author Leo Tolstoy wrote, "is not a curse but the glad business of life." It was from Tolstoy's ideas that Gandhi took his thoughts on bread labor. Here are the results, according to Tolstoy, that we would get from doing for ourselves and for which cooking can be a start:

> "You will become a more cheerful, a healthier, a more alert, and a better [person]."
>
> "You will learn to know the real life, from which you have hidden yourself, or which has been hidden from you."
>
> "If you possess a conscience, it will cease to suffer as it now suffers when it gazes upon the toil of others, the significance of which we, through ignorance, either always exaggerate or depreciate."
>
> "You will constantly experience a glad consciousness that, with every day, you are doing more and more to satisfy the demands of your conscience."

> "You will escape from that fearful position of such an accumulation of evil heaped upon your life that there exists no possibility of doing good to people."
>
> "You will experience the joy of living in freedom, with the possibility of good."
>
> "You will break a window—an opening into the domain of the moral world which has been closed to you."

We don't have to be in a fight with food. In fact, it is all pretty simple. Eat more plants. Eat less meat. Avoid things that are processed and advertised. Figure out where the food is grown and ask yourself if you agree with the values of the farmer. Break your bread with others and don't eat alone. Learn to actually cook for yourself.

It can take some time to change these habits, but if you take the ukulele approach and make just one change at a time, it will come easily. When you are done, you get to feel that the food given to you by the work of others—perhaps the most fundamental symbol of our interconnectedness—is in line with both how you want to be in the world and how you want to treat your own body.

14

OWN WHAT REALLY MAKES YOU HAPPY

The Difference Between Having Nice Things and Having Great Times

AT A TIME IN MY LIFE WHEN I STILL CHASED AFTER STUFF I sit on the bow of a friend's lovely sailboat, a mile off the coast of the island of Martha's Vineyard, with my legs hanging overboard. As the boat pushes through each swell in the Vineyard Sound, my feet plunge in and out of the water. I love it so much.

Yet I feel sad.

My mind keeps turning to the fact that I do not have a boat of my own. I keep thinking about how I want my life to be more like this moment. I want to sail more. What I need, I think, is to own a boat.

So next summer, I get one—an old secondhand one that needs a lot of work. That means what I also get is a month of weekends working my butt off to strip and refinish the wooden trim and to paint the boat's hull. Then there is the time and money I spend on repairs once the boat is in the water. And the stressing to spend more time on the boat than comes easily because, after all that effort, I feel I ought to.

What I had wanted was to re-create the amazing experience of sailing I'd had and to create new experiences of sailing that I imagined I could have. What I got that summer was a few nice-enough sailing experiences outweighed by the much less pleasant experience of owning a boat.

AT A NEW YORK CITY ELEMENTARY SCHOOL

I'm in a classroom looking at a crew of sweet-faced third graders. I'm talking about how the production of so much stuff in the world is straining our habitat's resources. I mention how the sad thing is that we may be wrecking our habitat to produce a bunch of possessions that don't even really make us happy.

As an example, I ask the kids, "Raise your hand if you have the experience of really, really wanting a toy or thing and nagging your parents and being really sad about not having it, and then getting it, and then being really disappointed because it just wasn't that fun and didn't make you as happy as you thought it would."

Every single one of the kids raises a hand. So do the two teachers. Then one child pipes up and says, "Actually, I feel that way almost every time I get a new thing."

Do this:

> Write down the *ten most expensive material things* you have bought over the last five years. This could be anything from a house to a car to expensive jewelry.
>
> Now make another list of the *ten experiences that have added the most value to your life.* Some might be things you bought, yes. But be honest. Because some might also be things like a regular Sunday walk in the woods with a friend. Or being on an Ultimate Frisbee team. Or listening to music. Or appreciating natural beauty.
>
> Now put a check mark next to the things on your list of

expensive stuff that actually helped you to have the most valuable experiences on your second list.

With your list in hand, ponder this: How much do things you've worked yourself into the ground to buy actually contribute to your life's value?

Relationship to Stuff Tip #1: Savoring the Bird in the Hand

A bird in the hand is better than two birds in the bush because you already have it. A good experience you are already having is better than a good experience you hope to have but may not.

Instead of buying a lot of stuff in the hopes of creating good experiences—the standard life approach to possessions—a lifequester knows it takes a lot less effort and is a lot more effective to ensure that you are getting the full jolt out of the good experiences you are already having.

According to social psychologist Fred Bryant, author of *Savoring: A New Model of Positive Experience,* these are some proven methods for maximizing the joy of—savoring—your good experiences or even making your average experiences better:

Share with another that you're feeling good. It actually boosts the experience further.

Take a mental photograph. Listing the elements of the experience that you enjoy makes you enjoy it even more.

Congratulate yourself for having the experience. It will give you the nice feeling that there is more to come since you are the one who helped create the experience you are having.

Compare it to a bad experience. This gives you a reference point to help you realize just how good your experience really is.

Remind yourself how quickly the experience will end. This helps you to stay in it and relish it.

Two Fixes to Improve Our Relationship to Stuff

1. BURST THE FALSE EXPECTATION BUBBLE

The reason I was so disappointed with the sailboat and the eight-year-olds were so disappointed with their toys was because neither the boat nor the toys brought the hoped-for experience. It wasn't because there was something wrong with the boat or the toys per se. It was because there is something wrong with our expectations of our standard relationship to material objects.

In practice, good times are only partly facilitated by things like boats and toys. Most *peak experiences*, as the psychologists call them, occur because of a conspiracy of circumstances including our state of mind, the novelty of the experience, the people we are with, and how much we lose ourselves in the experience. Yet in the media and in advertisements, we see people having exciting experiences or laughing or joking or being paid attention to *entirely* because of something they own.

In the world of media and advertising, dads feel closer to their kids when they bond over a new car. Romantic partners feel more loved when they get a gift that includes a diamond. Teenagers feel more popular because they have the right sneakers.

On the other hand, in the real world, kids and dads feel close only when they spend time together. Romantic partners feel loved when their lovers actually love them. Teenagers feel liked when they hang out with kids who are generous with their affection.

There is a mismatch between what we are taught to expect from stuff and what we can actually get from it. But when we are disappointed, we don't realize that our expectations of the very paradigm of acquiring and owning are wrong.

Instead, we tend to think that the problem lies with *what* we have acquired. The car is not fun enough. The diamond is not big enough. The sneakers are not cool enough. So we go back to work to earn more money to buy another thing, and hope that this time it will be the right one.

To break that cycle, we have to burst the false expectation bubble and be clear about what stuff can and cannot bring so we can invest life energy where it really counts.

2. ASPIRE TO AVOID OWNERSHIP

Ownership is a good paradigm for things we use all the time and want no one else to use—like underwear, socks, and toothbrushes. But consider this:

In the case of my loving sailing but hating boat owning, maybe what I really wanted was not to have a boat but to sail on other people's boats more often.

In the case of those third-grade kids and their presents, maybe what they really wanted was not to own a new toy but to try out some new toy a couple of times and then move on to the next thing.

Our standard relationship with stuff—long-term ownership—makes the stuff less effective in bringing us happiness. Why do we each own our own vacuum cleaner when they spend most of their lives in closet space we have to pay for? A book spends most of its time on the bookshelf we had to buy for it. A car spends most of its time in the parking space we pay for.

Meanwhile, manufacturers exploit faraway labor and waste our planetary resources to create all this stuff we barely like and hardly use.

Might it be better if, instead of treating the world as one big shopping mall, we found a way to treat it as one massive public library?

What we are going to talk about in this chapter is how to have a relationship with stuff that is better for us and better for the world, based on:

1. Having realistic expectations of stuff—that is, identifying when it helps and when it hinders our lives and the world.

2. Seeing that the effort required to own and maintain stuff often outweighs the good experience the stuff can bring.
3. Finding ways to extract the good from stuff (by using it) while leaving behind the bad (by not owning it).

Relationship to Stuff Tip #2: Defining the Experience You Want

The next time you consider buying something, take the time to carefully define the experience you are really trying to create.

If you find yourself wanting an expensive present from your partner, what is the real reason? Do you need proof that your partner loves you? Do you want some way of proving to other people that your partner loves you? Will the gift you have in mind accomplish this? Or could you save your partner the money and find a more direct and meaningful way to feel love with your partner?

Take the time to really understand what experience you want to create and ask yourself how to get it. Wanting or owning stuff may not actually be wrong. But doing so may not be the best or most direct route to creating the experience you want.

What Would You Save in a Fire?

If you think about it, by investing so much time and energy into acquiring our stuff, we may not be risking our lives, but we are using them up.

So what is worth using up your life for? I asked members of my social networks what they might save in a fire. Almost nothing could be bought from a store:

PHOTOGRAPHS AND JOURNALS

The pictures and journals I have will be added to and passed to my boys . . . and I pray they do the same for their kids.

—Leah Ritchie Andrews, Sequim, Washington

FAMILY JEWELRY

If I was to keep one possession, it would be the ring from my grandmother, Mammy, which reminds me of her as I wear it daily.

—Destin Lane, Brooklyn, New York

FAMILY MEMENTOS

My maternal grandmother gave me a ceramic piggy bank. I love that piggy bank, and every time we move I wrap it up with such care.

—Kathleen Calhoun-Chandler, Abington, Pennsylvania

COMPUTER/SMARTPHONE/IPAD

My laptop would be the only real prized possession because it contains all the information I've learned, hope to learn, and find valuable.

—Zach Folwick, Seattle, Washington

MARKS OF ACHIEVEMENT

I tend to use the same coffee mug over and over . . . and guess what? It was an award for a 12K I ran.

—Lydia Nemeth Odell, Brevard, North Carolina

BOOKS

I can't think of anything other than my book collection that comes close to representing my history, my interests, dreams, opinions, and loves.

—Jilliann Law Grygier, St. Louis, Missouri

MUSICAL INSTRUMENTS AND SEWING MACHINES

Our piano. It's not a family heirloom. But it's in the middle of the house and there's almost always one of us playing.

—Tessa Blake, Venice, California

HEIRLOOM FURNITURE

The dining table our family of nine ate at every night and then my own children grew up eating at, handmade by my deceased father when he was sixteen using trees he cut down.

—Kellie A. Parsons, Southern Pines, North Carolina

How to Get Into Heaven

There is a biblical story about ownership and life satisfaction known as "the parable of the rich young man." The story ends with Jesus telling the rich young man to sell all his things, give the money to the poor, and then join in with Jesus's disciples. The young man can't bring himself to do it and leaves, after which Jesus says to his followers, "It's easier for a camel to go through the eye of a needle than for a rich man to get into heaven."

You've probably heard that story and expression, and maybe, like me and many others, you made the mistake of taking it literally. Like, dude, if you have desires and live on this earth as a human being, you can't get into heaven? You can't be rich? Having and owning stuff somehow makes you bad? You're supposed to take some sort of vow of poverty? Damned if you want and don't get; damned if you want and do get?

But before we dismiss the story, let me interpret it in my words to hopefully cast a little more light on the kind of relationship to stuff that actually helps us and the world. If a story about Jesus strikes too religious of a tone for you, bear with me. Such stories deeply influence the culture we live in, so they influence our lives whether we believe them or not. Be patient. Unpacking the story of the rich young man offers wisdom and freedom for lifequesters, religious or not.

Anyway.

A rich young man came to Jesus and asked, "What must I do to live in Truth?" (*Truth* is my word. The Bible says, "What must I do to enter Eternal Life?" But no one who wrote the Bible was there, so who knows what words the rich young man actually used? Plus, there are real arguments to suggest that Eternal Life and Truth are the same thing, but that's a discussion for a whole other book.)

Now, before we go any further, let's take a step back. During Jesus's time as a spiritual teacher, many people came to him with practical, emotional, and spiritual problems, looking for advice.

They were not all rich like our young man. Many were poor. Some were sick. Some had problems with family.

There was a reason that they sought out a spiritual teacher for their problems rather than, say, a doctor or a lawyer or a banker. People who go to spiritual teachers go because they seek to relate rightly to their lives. They are lifequesters, of a sort. Among other reasons, seekers look for guidance from spiritual teachers when they feel distracted from their own Truths by the conditions of their lives.

That was likely the case for our rich young man. He must have felt somehow discontent to begin with, since he bothered to look for a spiritual teacher. We can assume he was a seeker who had yet to find a meaningful relationship with the conditions of his life. For these reasons, the rich young man sought out the nearest spiritual teacher who people seemed to think had a lot of answers—Jesus.

Then he asked his question. "What must I do to live in Truth?"

As the Bible tells the story, one might think that Jesus gave him a quick, almost impatient answer and that the interaction was soon over. That is because the scripture gives only the very bare bones of the interaction.

Since, generally, students and spiritual teachers have longer conversations, where the teacher asks questions and the student fills in some background, the actual encounter between the rich young man and Jesus was probably longer than the story implies. Perhaps the rich young man painted a more in-depth picture of his circumstances for Jesus. Maybe he said, "I should feel lucky. I have a lot of money and things. I have lands and cattle and servants and subjects. But it's like there is no treasure at the end of the rainbow. I'm just not finding any meaning. What can I do to find my Truth?"

Jesus then, according to the biblical version, gave a teaching to the rich young man that almost every spiritual teacher—from Moses to Buddha to Gandhi—gives to nearly every beginning student. That, to have a peaceful mind, you have to begin by building a peaceful life. So Jesus said to our malcontent, "You know the commandments: do not murder, do not commit adultery, do not steal, do not bear false witness, do not defraud, honor your father and mother."

The commandments are shared in one form or another by nearly all world religions (the Buddhists call their version the "precepts"). The behavioral guidelines form a kind of Spiritual Life 101. If you don't steal or kill or lie or hurt people with your lust, you have less conflict with those around you and your life becomes calmer. Then your mind, too, becomes calmer, and you are better able to hear your Truth—another reason why you must do right by the world to do right by yourself.

But as it turns out, our discontented rich man already followed this rudimentary teaching. He must have been quite a sensitive soul because he told Jesus he had kept the commandments since his childhood. But he still felt empty. In despair, he said to Jesus, "What am I missing?"

At this stage, there may again have been more conversation between Jesus and the young man than is reported in the scripture. Jesus might have said, for example, *Well, it's good for your Spirit that you keep the commandments, but have you also tried meditation?* Or: *That is good about the commandments, but how much service do you do for other people?*

Jesus may have offered a whole string of suggestions: *Maybe you need to take a break. Maybe you are stuck in doing things you don't want to do. Or maybe your life and your values don't line up. Maybe you need some time and space to reexamine or just to breathe. Maybe you need to go on a lifequest.*

The discontented rich young man probably resisted—because we all do. Often we know, deep inside, what we must do, but we feel trapped by our life circumstances (really meaning we are scared to change them). The rich young man may have offered excuses. *I'd like to meditate, but I have to meet with accountants first thing in the morning.* Or: *I'd love to help people more, but I really need that time to count my cows.* Or: *I just don't have time to take a break. Who would take care of my lands and possessions?*

Perhaps during this longer conversation, it became clear to Jesus that the rich young man was too attached to his belongings. He allowed himself to believe that his well-being and good experience of life came from what he owned. That erroneous belief kept him in his habitual behavior and stopped him from doing what was clearly necessary to follow his Truth—to prioritize things that were more likely to lead him to a fulfilling life than his possessions apparently were.

The rich young man didn't view property as just one of many means of potentially helping himself and others. He didn't see his property as a potential way of doing service. He didn't see possessions as a tool to be sometimes picked up and sometimes put down. He had come to see owning and maintaining his property as the source of his security and an end in itself.

So Jesus, seeing this, may well have said to the young man, *You need to stop running your life as though the point of it is to take care of your stuff. You are barking up the wrong tree. If you are serious about finding your Truth, you need to leave that stuff behind for a while. You need the time and space and freedom to figure out who you really are without being overwhelmed by your obligations to all your things. Let them go for a while. Come travel and study with me.*

Or, as the Bible says, "Go, sell everything you have and give it to the poor. Then you'll have treasure in heaven"—possibly meaning that such a charitable act will give you a boost and make you feel more meaningful. "Then follow me."

That might sound radical, but all Jesus was really saying was that the rich young man should go on a retreat. He didn't say, "You have to hate possessions forever. You can never own anything again." Nor did he say that owning things or being rich was wrong or bad.

He simply saw that the rich young man's relationship to ownership stood in the way of his Truth at that moment. His possessions were taking all his time and energy and focus. So Jesus suggested that the rich young man leave off possessions for a while and instead go on his lifequest. Find his Truth. Search.

But the rich young man wasn't ready to go on a retreat. At least on this day—and for all we know, he changed his mind the next day—he was scared that once he let go of his stuff, he would never have it again. He feared he would never be safe because he didn't know where his True safety came from (remember the problem of not being self-reliant we discussed on page 195?).

The rich young man couldn't see that the stuff as well as the effort and emotional energy he put into keeping it were standing in the way of his Becoming. He couldn't see that there are times when our stuff helps us and times when it hinders us. He could not bring himself to sell his estates. Instead, his face clouded over. He just turned around and left.

When he had gone, Jesus looked at his followers and said sadly, according to scripture, "How hard it is for the rich to enter the kingdom of God." He might just as easily have said, in the language we have been using: *How hard it is for people to let go of their possessions, even when those possessions stand in the way of their following Truth and becoming the people they want to be and living the life they yearn for.*

He didn't say the rich were bad or wrong. Rather, he was expressing his compassion for the challenge faced by the discontented rich young man in letting go of his attachments. It takes a lot of courage to let go of the standard life approach, even for a person for whom it is failing. But when you have a lot of property and you are "successful" on the outside, it can be hard to let go of your attachment to property, even if you feel meaningless on the inside.

A few Bible verses later, Jesus reassures his followers that anyone who leaves behind his riches to follow a life that is truly authentic—anyone who acts out of courage and faith in his True Self instead of fear—will be rewarded a hundredfold in his life. Everything that is important will come to you as long as everything that is important is what you really want.

People who follow their Truth and give their energy to what actually makes their lives and the world better, and are not distracted by the superficial desires that don't, will get into "heaven," as it were. Knowing that the rich young man couldn't trust in such an assurance is why Jesus must have felt so bad about his shortsightedness.

When Jesus said, "It is easier for a camel to go through the eye of a needle than for a rich man to enter the kingdom of God," it was not a judgment or a condemnation of people with possessions but an expression of compassion and empathy for the challenge of letting go of what you erroneously think makes your life worthwhile.

None of this is to say that making use of material things is bad or wrong. A roof to keep us dry, a pot to cook in, a surfboard to surf on—the use of such things helps us feel safe and enjoy life. When we feel safe and happy, that is when we have the energy and mental capacity to help our worlds.

When we decide that the use of things is bad and deny that we sometimes need them, we make ourselves just as miserable and unavailable to others. Having the use of too few things can diminish our ability to follow our Truths just as much as chasing after the ownership of things can. Antimaterialism can be yet another calcified standard life approach, especially when it is an angry reaction to the materialism of others and lacks love and compassion.

Dogma—both materialist and antimaterialist—doesn't work. There are no rules you can follow that will lead you along a path to the Good Life. There is no belief system—no set of stories—that replaces being awake and responsive. You must follow the path that emerges for you in each moment—what Buddha called the Middle Way.

The question we have to ask then is, what is the middle way in our relationship to things that is most effective in leading to Good Lives for us and our world?

Relationship to Stuff Tip #3: How to Not Buy Things by Mistake

Don't make shopping a social activity: Shopping has become one of America's top two pastimes (watching TV still beats it). It has become a social activity. And what happens is that you end up buying stuff you don't really need when you actually just want to spend time with your friends. So just do something else. Plan to cook together, craft together, exercise together, or play games together, but don't plan to shop together.

Leave your credit cards at home: All the studies show that we don't feel we are spending our money when we use credit cards, so we are more likely to buy things we can't afford and don't really want or need. One study out of Purdue University showed that people who merely *saw a credit card logo* were willing to pay three times more for a piece of clothing than people who didn't see it. In another study, MBA students at MIT were willing to pay more than twice as much for sports tickets when they could use a credit card instead of cash. *Fill your wallet only with what cash you can afford to spend and leave credit cards at home.*

Never shop on your mobile device: Research says we spend more when we shop on our mobile devices. Why? Partly because online retailers gamify shopping and take advantage of those boring commuting or waiting-in-line moments. *Read, knit, solve puzzles, or listen to music when you are bored, but don't shop on your mobile device.*

But Isn't There Something in Human DNA That Forces Us to Shop?

One thing you'll hear is that humans are programmed by their biology to shop and that our need and desire are insatiable. As the story goes, back when humans lived in the jungle, resources were scarce. When we came across them, we hoarded them, because who knew when we would find more? That instinct, the story goes, is what makes us shop. It is just in our nature.

That is a great story for justifying the status quo. The only problem is that it doesn't actually have the benefit of being true. If it were, looking at history, we would see that human beings would always have shopped when they could and nothing would have to be done to persuade us to do so, right? But that isn't what history tells us.

About three hundred years ago, the prevailing economic question was: How can society produce what everyone needs? Back then, before the Industrial Revolution brought along factories and mass production, people made everything from clothes to furniture by hand. Pots and pans, for example, were made one at a time by blacksmiths and metalworkers.

For that reason, you couldn't just pop around the corner and expect to buy a pot or pan whenever you wanted. In fact, when a traveling merchant—known as a tinker—rode into town in his horse-drawn wagon and sold you a new frying pan, oftentimes he'd accept the broken one you wanted to replace as partial payment. He would repair it and sell it again.

Because things were made by hand, and because that made them harder to come by, people respected and valued those things. Even if they could, no one went shopping just for fun and no one hoarded stuff—or thought of shopping as part of human nature. Quite the contrary. Buying stuff you didn't need just made it hard for other people to get stuff they did need and proved you weren't being careful with your own resources. Wasting was frowned upon. Being thrifty was virtuous.

Then came the Industrial Revolution. Instead of those pots and pans being made one at a time by a metalworker, factories churned out thousands at a time. You didn't even have to wait for the tinker's next visit to get stuff. You could go to one of the new "department stores"—the old-fashioned version of the mall—or order from the Bible-thick Sears, Roebuck & Co. catalog—the old-fashioned version of Amazon.com.

Still, the shopping frenzy did not begin there. Once you had a Sunday dress, why would you need two? If your parents gave you your grandmother's silverware, you felt proud to display your family heritage at every meal. Why would you buy more? People still didn't buy stuff just because they could. They bought only what they needed. They still repaired and made do. Neither our DNA nor our biological heritage forced us to hoard.

Then, back in the 1940s and 1950s, thanks to further mechanization and other new production methods, industrialists worried that the public's needs would soon be saturated and that factories would come to a grinding halt. Where once the economic question was how to produce what everyone needed, now the question became how to get everyone to want what society produced—even if they didn't need it.

Marketers had to figure out how to make what they were selling look better and more valuable than what people already had. The answer the marketers came up with was the use of advertising to convince customers that the new product would meet a compelling personal need the old possession didn't, *even though it wouldn't.*

The real beauty of this system, from the industrialists' point of view, was that, since the product would never really scratch the itch it was supposed to, satisfaction would remain perpetually just out of people's reach. Dissatisfaction, by contrast, would be part of the process. Customers would be caught in a kind of purchasing hopscotch. *Well, that didn't work, maybe I'll try buying this. Nope. Maybe I'll try this. Nope. Maybe this? Nope.* In this way, society's needs would never be saturated, as the industrialists feared, and the economic wheels would keep turning forever.

As an example, consider one 2014 car advertisement. It shows a kid who is upset because the toy his dad bought him doesn't work. The dad is distraught. The emotional effect of the scene on today's time-starved dads watching the ad is to activate their fears of never having the chance to truly bond with their children.

But in the next scene in the ad, our TV dad cheers up his child by having a great driving adventure. Real-life dads at home watching the story from their couch feel a sense of relief on behalf of the dad in the advertisement. They can't help having a good feeling about the car, since it seems to bring the same happy ending between father and child that they hope to have.

That's how many ads work: they activate a fear of not getting a really important need met and then associate the product with the pleasant feeling of being released from the fear. Our car ad associates a certain brand of automobile with the happiness a love-starved modern dad feels when he finally bonds with his kid. Who doesn't want that?

Except, of course, the new car won't bond the father and the son, the new outfit won't get us a job, and the new cell phone won't make anyone think we are cool.

It is hard to believe we fall for these ads until you consider that first, the ads play on our deepest fears—ranging from our mortality to our social acceptability. This activates our fight-or-flight response and short-circuits our ability to reason. Second, even if we could effectively evaluate an ad's message, we don't have the time, since each of us is buffeted with more than ten thousand ad messages every day. And finally, the ad messaging is replayed and reinforced through our interactions with media and with everyone we know who has been force-fed the same messaging.

All those ads, pumped straight into our homes and offices and lives for the last seventy years by our TVs and computers and handheld devices, add up to one huge propaganda campaign that says that your deepest fears could come true at any moment. The good news, according to the propaganda campaign, is that if you just buy the coolest and most fashionable and most timely thing, at least there is a chance you will be successful and better loved and you will have a better life.

Even if, for a moment, you wake up to the craziness of the message, do you really want to be the one father, mother, high school kid, friend, or lover who doesn't participate in buying the stuff? How much pressure do we all feel from everyone who keeps telling us how cool this pair of sneakers or that cell phone is? How unacceptable might we feel if we disagree?

How much courage must a rich young man—or woman, of course—have to break away from this standard life approach?

We started, in this section, with the question of whether it is human nature to want to shop, whether it is somehow coded into our DNA. What we've seen is that, according to history, there have been times when people didn't shop and times—like now—when we do.

When we didn't shop a lot, it was because hoarding stuff was not considered acceptable by one's friends, families, and peers. Now that we do shop a lot, it is partly because we believe it makes us more desirable in the eyes of our friends, families, and peers.

In other words, it is not the shopping or the not shopping that is important to us and part of our DNA. What is important to us is feeling like we are valued by the group or groups of people whom we value. What's important to us is that we have a place to love and be loved, that we fit in, that we can feel safe.

Relationship to Stuff Tip #4: How to Decide if It Is Worth Buying

Next time you are thinking of making an optional purchase, don't ask whether you can afford it; ask whether it will actually bring you what you hope it will. The way to do that is, first, to ask yourself what you really hope to get from it, and second, to see if people who already own it got what you hoped for.

Don't ask them if it was worth it, because science shows that people are highly motivated to justify their own choices. Just watch and see if having the product made them truly happier in the way you imagine the product might. This will help you to know if a product actually delivers what you hope to get from it.

Nine Ways to Be "Acceptable" Without Buying Anything

Most advertising today functions by convincing you that you are less adequate than those around you. Your body, your parenting, your work, your looks, your conversation—none of it is as good as other people's. The ads prey on our insecurities and then tell us a product will fix them. Except it won't.

Here are nine insecurities advertisers exploit to convince you that you are inferior, together with ways to fix that insecurity without buying anything. While you are probably doing as well as anyone else is in real life, it helps to remember the people in the ads are fake.

1. When ads make you think you should cook more:
They are selling: A juicer, an ice cream machine, a bread machine, a sandwich maker.
Bonnie Curtis from Lakeville, Massachusetts, fell for the advertising hype, bought a food processor, and regretted it. "It's a pain to clean and a good knife and cutting board work just as well."
Do this instead: If you really think you should cook more, all you have to do is . . . cook more (see pages 183 to 192 for tips).

2. When ads make you worry your life is out of control:
They are selling: Closet dividers and boxes and shelves and "organization systems."
Kristy Everson from London, UK, tried buying a bunch of stuff from the Container Store. Then she realized, "I didn't need stuff to keep my stuff in. I just needed to donate a lot of the stuff I already had."
Do this instead: The best way to organize your stuff is to have less of it—purge.

3. When ads make you feel you are less handy than you should be:
They are selling: Electric tools and tool kits.
Guess how much the average electric drill is used in its lifetime? Between six and thirteen minutes. And guess how many U.S. households already own an electric drill? Fifty million. So if you need a tool, chances are that there is already one collecting dust in your neighbor's garage or basement.
Do this instead: Being handier doesn't mean owning more tools. It means learning how to use them. Save the money and take a class.

4. When ads make you worried about parenting your newborn:
They are selling: The baby gift registry at baby supply chain stores.
Worried you won't know what to do when your new baby comes? Think maybe you should have lots of gear stashed, just in case? Aimee Chou from Seattle, Washington, says, "We never used about 80 percent of the new stuff we requested on our baby registry."
Do this instead: Don't trust baby supply chain stores for parenting tips. Trust your heart and spend time learning from other parents you really admire.

5. When ads make you feel uncreative and boring:
They are selling: Sewing machines, electric potter's wheels, and all sorts of kits.
Marketers convince us we need their stuff to express ourselves when really we just need some time. Melissa Kay Cervone from Seminole, Oklahoma, says, "I bought a knitting loom and I haven't used it once yet. I should have just worked on my crocheting instead."
Do this instead: To be creative, set aside even half an hour a week to draw, write, sing, sew, or dance. Once you start, the time you find will grow.

6. When ads make you scared you're unattractive:
They are selling: Clothes, fashion accessories, personal products, and makeup.
If you are not stylish, won't your buying clothes just mean you'll have more unstylish crap in your closet? And if you are stylish already, why do you need new clothes when your closet must already be full of stylish stuff?
Do this instead: Scavenge your own closet and combine your best items into your best outfits, maybe with the help of someone whose style you admire.

7. When ads trigger your nostalgia about an old sport:
They are selling: Tennis racquets, skis, and camping gear.
You don't need your own equipment until you're sure you will stick with the sport.
Do this instead: If you want to get back into a sport, commit to doing it not by buying something but by making a date to do it with friends—and borrow their extra equipment.

8. When ads make you hate your body:
They are selling: Gym memberships, exercise equipment, and diets.
The diet industry has us convinced that we need some special sort of expertise or equipment to lose weight and be healthy. Robbin Vokes Lenardon from Prescott, Arizona, bought a second bike trainer because her "hubby thought we might exercise together. No way!!"
Do this instead: You don't need to buy anything to exercise or diet. Start exercising by going for fast walks and lose weight by eating fewer calories than you burn.

9. When ads make you feel like you rock and don't need to change a thing:
They are selling: Nothing. Ads will never say this. Their job is to make you feel bad about yourself, not good.
There is only one place where people cook more than you, love

better than you, have sexier bodies, and all the rest. That is in the media—on the Internet, in the movies, and on TV.

Do this instead: *Slow down!* Evaluate the con. Do the people in the ads really exist? Look at your neighbors, your friends, your colleagues. Do you really in your heart of hearts believe you are less than they are? Aside from the fact that you are stressed and harried and feel a bit insecure—just like everyone else—when you refer to your Truth and not to TV or the Internet, wouldn't you say you are doing pretty well without more junk?

What We Pay for Stuff

Pounds of raw materials used to make the average world citizen's stuff every day: 50

Pounds of raw materials used to make the average American's stuff every day: 362

Number of highly populated countries using more raw materials for their stuff than America: 0

Number of American households with storage units full of stuff: 10.9 million

Percentage of storage renters who say it's because stuff won't fit at home: 50

Number of times bigger the average U.S. house is compared to fifty years ago: 2

Average annual cost of a 100-square-foot storage unit: $1,380

Credit debt of the average U.S. household (largely from buying stuff to put in storage): $15,480

Annual interest paid on that amount at average 2014 interest rate (21 percent): $3,250

Approximate growth in self-reported happiness in the United States since the 1950s: 0

Countries whose citizens reported feeling happier than American citizens: 32

Countries ranking higher than the USA for effectively using raw materials to create happiness: 104

Number of planets required for all of humanity to live like Americans: 4.1

Number of reasons people should want to consume like Americans: Um. Zero?

Squeezing the Juice from Stuff and Leaving Behind the Pith

What we want, in general, is to have a good experience of life. We want our minutes and hours to be fun and absorbing and love-filled and meaningful and pleasant.

Most of us want that good experience of life for ourselves and others. The lifequester, in particular, recognizes that in many ways, the extent to which any of us can have a good experience of life depends on whether other people get it to.

For these reasons, the value of our relationship to a material item consists entirely in the totality of the experience it brings, both to ourselves and others. To what extent does the way we relate to a thing improve our and everyone else's minutes and hours, and to what extent does it degrade them?

Mathematically, we could put it like this:

True value of our relationship to a thing = Total good experience it brings us and the world - Total bad experience it brings us and the world

We have spoken a lot about how our standard approach to stuff—

"get and own more of it"—brings more bad experiences to ourselves and the world than we realize. On the other hand, while there is a lot we can't get from stuff, there is also a lot we can. For example, you can't go sailing without a sailboat. Can't get dressed without clothes. Can't go on a bike ride without a bike. Can't knit without knitting needles and yarn. Can't write without a pen or typewriter or computer. Can't stay out of the rain without a roof.

The lifequester is not materialistic and not ascetic. The lifequester follows Truth. When we discussed the rich young man, we looked at whether our relationship to stuff supports what is important to us and our fellows or stands in the way of it.

So the question is: How do we make it so we maximize the service and experience stuff offers without incurring the personal, societal, and planetary costs of continually buying and owning it? How do we begin to have the kind of relationship to things that maximizes their value in our lives?

1. ENJOY STUFF WITHOUT OWNING IT

One way to start this process is to remember that stuff itself, when you separate it from the service it provides, is actually a burden. We have to buy it, maintain it, store it, and dispose of it. We keep paying for our stuff, in money and effort, even after we buy it.

Recent studies show that even the ownership of a home or second home—something we are all supposed to want—does not make us happier than renting. That's because when we own a house we are too busy arranging to have the boiler replaced and scrimping on vacations to pay for it to fully enjoy other aspects of life.

But you can still enjoy a country house by renting it—you'll pay only for the time you use it. Nor will you be stuck with the house the following summer when it ceases to excite you. Renting or borrowing (you can find lots of sharing and renting websites online) allows us to enjoy stuff without the downside of owning it.

Say, like me, you want to sail more. Don't commit yourself to the $7,000 average yearly cost of purchasing and maintaining a small sailboat. Instead, go to a yacht club on the East Coast of the United States in the fall and scrounge around for a crew position on one of the boats going to the Caribbean for the winter.

How can you afford to take a whole month off to sail? By not taking that standard life approach that has you paying to own sailboats and country houses. The value in stuff is in using and enjoying it, not owning it.

2. LIBERATE WHAT YOU DON'T USE

Americans use 80 percent of what they own less than once a month. What sense does it make to own something you don't even use? First you buy it, then it takes up space, then you clean and dust it, then you fix it, then you bury it under other stuff, then you buy another one because you can't find it. Does it even add to your life? How many resources and how much labor would be saved if we shared these things instead of owning them?

Look around at your things. How many things do you own that someone else might use? How much furniture; how many tools? How many of them would you actually miss? How much space would you have if you got rid of them? Right now, find five things to donate or give to friends.

Or if an item you don't use is too valuable to give away, then go to one of the many sharing websites and figure out how to rent it out or loan it to neighbors so they don't buy and waste the same thing.

3. WANT WHAT YOU HAVE MORE THAN WHAT YOU DON'T

Why would you want to play squash when you have three tennis racquets? Why buy a specialized rotisserie oven when you already have a regular oven? Why buy a ukulele when you already have a guitar? There is a special happiness that comes with savoring and treasuring and reusing the stuff you already own.

I've had the same bike for fifteen years. It was given to me by a friend after my previous bike was stolen. Once I got hit by a car while riding it. Afterward, it needed work, so a guy named Joe who worked at Recycle-A-Bicycle, a nonprofit that trains inner-city youth in bike repair, fixed it up. Bella, my daughter, has sung so many songs from the back of that bike as we rode along. Where would all those memories go if I got a new bike?

Suppose you need a book for vacation. You could go buy another one at the store. But maybe you have one on your bedroom floor that somebody you love has been telling you to read. Wouldn't taking that book along be like bringing part of your loved one on your trip? Or maybe you have a book that changed you when you were younger. Wouldn't it be lovely to reread it?

Repair, repurpose, reuse. When we squeeze more joy out of the stuff we already have, we find we don't want more of it.

4. GET WHAT YOU DON'T HAVE CHEAPER AND WITH LESS HASSLE

Looking at the diagram below, you can see that statistically speaking, products are liable to fail when they are either very new or very old. We know about the old part intuitively. Your computer hard drive tends to go kaput at exactly the time it has eight years of data on it.

Why Secondhand Stuff Can Be
More Reliable Than New

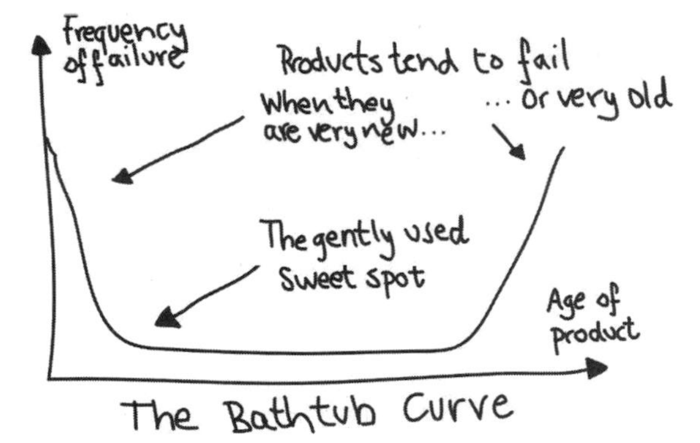

The Bathtub Curve

So most people think buying new means buying fail-proof. But think about it. When you have to return a product, doesn't that usually happen right away, not months down the line? If, for instance, your cell phone doesn't work, that's usually an immediate issue after purchase, not something that comes up three months later. This is why a nearly new, refurbished cell phone or computer is likely to be more reliable than a new one: the faults have been found and fixed.

Also, by buying secondhand, you save the resources required to buy something new.

Cool Stuff That Happens When You Buy Secondhand

Once I was looking at books at our local Goodwill. A family album caught my attention. When I opened the album there was a beautifully written message from a mother to her family. I couldn't believe that someone would give up this family history. I looked at the last names in the book to see if they lived near me. They did, so I called. The husband was so happy to hear about this book being found. His wife had passed away recently and some books had been donated and the album had accidentally been lost. He dropped by my house that night. It was wonderful to see how excited he was to get it back.

—Sara Diller Gallaugher, Columbus, Ohio

We went to a yard sale a few months back, and picked up some board games from this girl. Incidentally, she was moving to the house up our street. We ended up meeting for dinner, and playing the same games that we bought from her. It was so much fun building a beautiful relationship from a yard sale purchase.

—Anjana Sukumar, Boston, Massachusetts

One time I bought some baby items off of Craigslist and went to a house to pick them up. The seller, a father, came to the door in a smoking jacket with a pipe and a huge book of constitutional law. Yes, part of me was thinking I shouldn't go in. He seemed strange and it didn't feel safe, but damn it, this looked interesting. So I go in to wait and he takes me into their library, which was AWESOME. We are chatting and he is every bit as eccentric as I'd hoped and then he says, "See that sconce over there? Pull it." So again, yes, I am dumb, but I was like, "Hell yeah!" so I did and a wall moves and a secret stairway opens up. "It goes to our new theater room, you can check it out if you want." So yes, of course I did (stupidly, but come on, it's a SECRET ROOM). It was awesome. Anyway, looking back I'm sure it wasn't that bright but I just couldn't resist. I've never had a bad Craigslist experience and I've purchased and sold a ton. Each time I meet a nice person, exchange stories, sometimes even make a new friend.

—Lindsay Neely, Minneapolis, Minnesota

15

HOW TO GET AROUND AND WHERE TO STAY PUT

A Bean Counter's Approach to Improving Our World Relationships

Earlier, we talked about finding your first step on the lifequester's gentle path by looking for relationships to the world that have always nagged at you, that you have always wanted to change. Sometimes, living in ways we don't really like becomes such a habit that the nagging dies out.

Instead, we might have a sort of general malaise—some anxiety or depression—without knowing exactly what it is about. We just don't feel the way we should. In our quiet moments, when we look around at the world, we don't feel it is working the way it should, either. Yet we just can't tell what direction would move us away from the malaise and toward a solution to our own and the world's suffering.

In this case, we have to look for other external clues pointing us in our True Direction. One way to find those clues is to take what I call the bean counter's approach to improving world relationships.

To do this, to understand where your intuition might otherwise direct you if the noise weren't too loud, take an inventory—just as a bean counter might—of what you spend your life energy on. For our purposes, *the life energy you spend on something is the total of the time and the money that you dedicate to it* (we will talk more about life energy in chapter 27).

Now, since it is your *life energy* that you are spending, you should expect to get a decent value in return on investment for it (ROI, in the language of bean counters). In our case, the return on investment we are looking for is the feeling that our life energy investments move us toward a life well lived—the Good Life.

Taking an inventory of where we are wasting life energy—not getting security and meaning in proportion to what we spend—can help us determine which of our world relationships we should work on first. To use the bean counter's approach to choosing which of your world relationships need improving:

1. **Make a quick list of the things you spend your life energy on:** What do you spend your time doing? What do you spend your money on? It doesn't have to be a thorough list (you can do this in more detail later in chapter 27).

2. **Ask what need each item is supposed to fill:** Why are you spending life energy on this item? What is the promise it is supposed to fill?

   Go through your list. Regardless of whether an expenditure of life energy feels "necessary," ask yourself what return you get from each item in terms of security and meaning. Is the amount of true life satisfaction you're getting in line with the life energy you are giving up? Does the expenditure of time or money move you toward one of life's intrinsic values: spirituality, community, affiliation, self-acceptance, physical health, safety, or hedonistic pleasure?

3. **Notice which world relationships bring the highest and lowest ROIs:** Put plus signs by those world relationships that

bring high degrees of security and meaning. Put minus signs by the ones that bring low degrees.

4. Unpack the high and low ROI relationships: Ask yourself what elements of those relationships make them particularly helpful or unhelpful in terms of bringing security and meaning. Can you get the thing you need (identified in step 2) in a way that offers more security and meaning? Do you know other people who manage to do so?

5. Look for alternatives: Look at how other lifequesters are managing the relationships you want to change. How can you change the world relationships you have identified to increase your Good Life ROI?

How the Bean Counter's Approach Made Ralph Fall in Love with His Feet

Speaking of bean counters, remember our imaginary friend Ralph the accountant? He grew up with a passion for dance but decided it couldn't give him the safe life he wanted and abandoned it. He became an accountant and learned not to listen to the inner voices telling him he wanted something different.

For the longest time, he sat in traffic jams twice a day, on the way to and from work, and felt miserable. His answer was to put more life energy into his work to earn more money so that he could eventually afford a Mercedes with leather seats and a Bose stereo. Then the traffic jams would be bearable. Life would be good.

Let's assume, now, that one day at the office, Ralph has a mini-awakening. Ralph has always fantasized about the life of his boss's boss, who already has that Mercedes. Then, on this particular morning, Ralph overhears his boss's boss complaining about the traffic, just like Ralph does.

But wait a minute! Ralph thinks. The boss's boss already *has* a Mercedes with a Bose stereo. The boss's boss has exactly what Ralph thinks will make his life better but is complaining about the exact same thing Ralph complains about.

Suddenly Ralph wonders, if you are still going to be miserable in the traffic jams, is it really worth spending an extra $60,000 for a high-performance car that mostly won't be able to go more than twenty miles per hour because of the traffic? For just a minute, Ralph questions whether the route to the Good Life is really to throw more time and money—life energy—at his standard relationships to the world.

But how do you change when for so long you have been ignoring the quiet voice within? Well, being an accountant, Ralph is particularly intrigued by our bean counter's approach. How much life satisfaction bang does he get for his life energy buck? He decides to follow the directions we outlined above and makes his list of things he spends his life energy on. What can it hurt? It's just accountancy anyway.

Now, assuming Ralph is an average American, when he makes his list of spending, he sees that his second-largest expenditure, after housing, is his transportation. Indeed, from an article in *USA Today* he discovers that the average American spends nearly $10,000 a year owning and maintaining a car (parking *not* included), amounting to about 22 percent of the average American's wages.

This intrigues Ralph, since getting to and from work—his transportation, the thing the average American spends so much money on—is the thing that has been making him so miserable. *Surely,* Ralph thinks, *if it is my second-largest expenditure and it is making me miserable, I'm not getting the right Good Life ROI.*

When Ralph looks into it a little more, he discovers that, according to the *Harvard Health Publications* newsletter, average Americans like Ralph spend 101 minutes in their cars *every single day.* Add all that driving to a nine-hour workday and no wonder so many Americans, like Ralph, are too exhausted to do anything but spend another 170 minutes each day sprawled on the couch watching television or movies.

The thing is, all that money and time spent on car ownership is supposed to make your life better. But Ralph next discovers a whole bunch of studies showing that people who own cars are significantly *less healthy* than those who don't.

One study published in the journal *Obesity Research & Clinical Practice* showed that citizens of eight Chinese provinces who lived in a car-owning household were 80 percent more likely to be obese than those who didn't. Another study, published in the *Canadian Medical Association Journal,* showed that obesity among developing-world citizens increased by 400 percent after they bought cars. In fact, owning a car, a computer, and a TV corresponded to a nine-centimeter (three-and-a-half-inch) increase in waist size.

Meanwhile, back in the United States, thanks in part to the fact that we use wheels instead of feet to get around, Americans get an average of only nineteen minutes of exercise a day. As far as health is concerned, in other words, car ownership is the new tobacco.

On top of all that, it turns out that Ralph's and everyone else's cars are the cause of an entire one-fifth of all U.S. climate emissions, according to the Union of Concerned Scientists. Human rights

organizations also associate the oil and gas industries with all sorts of human rights violations; civil war; and land, water, and air contamination around the world.

In other words, using the bean counter's approach, Ralph is seeing that his 101 minutes spent in the car each day are both bad for him and bad for the world.

The good news, Ralph thinks, *is that at least a car saves me a lot of time by getting me everywhere faster and is a lot more pleasant.* Then he does a little more research and discovers that even *that* may very well not be true, at least in places where there are other transportation options (the sad news is that there are many places in the United States where there are *not*).

Let's say, for the sake of argument, that in the period when Ralph is mulling all this over, he also happens to notice, as he is sitting in the traffic jam, the commuter train whizzing past on one part of the highway and the bicyclists in the bike lane and the carpoolers and ride-sharers in the HOV lane whizzing past on the other.

How Slow Does Your Car Really Go?

As we've discussed, most of us spend 101 minutes each day in our cars, and the average American also spends 22 percent of her income on her car. Since the average American spent 1,788 hours working in 2013, according to the Organisation for Economic Co-operation and Development, that means she worked 393 hours (22 percent of 1,788 hours) paying for the car. That amounts to a grand total of 1,007 hours each year either in the car or paying for the car. The average miles driven by an American is 18,000. Divide those miles by the amount of time you spend either in the car or working to keep that car and you find that your overall average speed is a not-very-impressive 18 miles per hour.

Ralph begins to wonder if, on top of whizzing past him, the cyclists and train passengers and ride-sharers are actually having a better morning than he is. After all, the cyclists are getting exercise and the train passengers are reading or napping, and three out of four of the ride-sharers are avoiding the stress of being in the driver's seat.

When Ralph does the research, it turns out his guess is right. He finds a University of East Anglia study of eighteen thousand British commuters and a McGill University study of six thousand commuters that both show that solo drivers are the least happy of all the commuters. If Ralph looked at a bunch of research from around the world, here is a summary of what he might have found out about how to be a happy commuter:

> The quality of your commute can be as important to your overall life satisfaction as having a life partner or a child.
>
> The more you actively use your body and get exercise as part of your commute—cycling, walking, scooting, running, skating—the more satisfied you will be with it. In fact, the research shows you will be happier walking thirty minutes to the train station than driving the same distance in only five minutes. A longer commute using your body will make you feel better than a shorter commute using machines—despite the time difference.
>
> Like "active" commuters, mass transit users, train riders, and ride-sharers are all more satisfied with their commutes than people who drive alone.
>
> Mass transit users are most satisfied when they can settle down for a while—to read, to listen to music or podcasts, to chat, to nap—than if they have to transfer. So a half-hour train ride will probably feel better than a twenty-minute ride involving two trains or a train and a bus. A slightly longer trip involving fewer changes is more pleasant than a shorter trip with transfers.

Overall, the life-satisfaction research shows that driving back and forth to work—day in and day out—kind of sucks. You can't do anything but drive and you don't get to use your body. So for most of us, driving should be the transportation mode of last resort.

Using the bean counter's approach, Ralph discovered that driving a petroleum-powered box—even a Mercedes—to carry him around is neither good for the world nor good for him. When it comes to transportation, the best ROI for life energy spent is a pair of sneakers, a bike, or a scooter.

Thus, Ralph decides that the world relationship he wants to try to change first is the one involving transportation. This is how, of all people, Ralph the accountant first begins to fall in love with his feet.

Why NOT Biking Is More Dangerous Than Biking

One reason people give for not riding their bikes to work is that they feel it is dangerous—because of the air quality and the potential for crashes.

But the research shows that bicycle commuters live, on average, fourteen months longer than car commuters. This, according to studies on life expectancy, is because car drivers are sitting on their butts while bike riders are getting exercise.

In spite of the perceived dangers of biking, the chances of living longer *increase* when you ride a bike to work because the chances of dying of being sedentary—from a heart attack, stroke, etc.—go down.

Cool, right?

My Facebook Friends' Ten Tips for Happier Hoofing

When most of the culture is doing something one way, transitioning to doing it another way takes some time and patience. Most lifequesters with experience moving away from car use will tell you to give yourself time. You can't just wake up on Monday morning and expect to leave your car in the driveway. Give yourself a chance to figure it out and permission to go easy on yourself.

1. Find your way around: "Start by using the trip planners on the transit websites. These will tell you what routes to take, how long it will take, and if you have to walk part of the route. I've even learned to avoid using rental cars when I visit other cities in North America by looking up routes and schedules on the Web."—Heather Moran, Waterloo, Ontario

2. Combine your trips: "Even if you can't completely eliminate your car you can: (1) combine your errands to make fewer trips; (2) let friends and family know when you are going to the post office, library, etc., and travel together; (3) replace your old favorite spots with new, closer ones."—Joellen Gilchrist, Beverly Hills, Michigan

3. Live near the action: "My way of adapting to a no-vehicle lifestyle is to live in a central location, usually the downtown area of my current town. Often, there are small groceries there, cafés and restaurants as well as the hub of the transportation system."—Brenda Weitendorf, Brampton, Ontario

4. Have things delivered: "Plan a big online grocery shop once a month to have the bulky, heavy items delivered. Also, know the number of the local 'man with a van' who can pick up things that are too heavy to carry. It won't be free, but it's cheaper than owning a car."—Jane Micklethwaite, Nottingham, UK

5. Make it a game: "I gave myself mini-challenges. Take alternative transportation or walk or bike just for a week. See how it goes. Try different modes. Soon one week becomes two." —Nadine Arseneault, Toronto, Ontario

6. Plan for enjoyment: "If I can get a seat on the train, I read, but in case I don't, I always make sure I have plenty of good music and interesting podcasts to listen to."—Kristina Fukuda-Schmid, Culver City, California

7. Share a car: "One thing that helps people to transition away from car ownership is to enroll in a car-sharing service. Then you have access to a car when you need one."—Tanya Seaman, Philadelphia, Pennsylvania

8. Get a teacher: "Find a seasoned bicycle friend who can take you out on a Sunday and go over the route you will commute so that you feel comfortable doing it on your own."—Maureen Persico, San Francisco, California

9. Dress like a walker: "Carry a backpack instead of a purse—that makes it a lot easier to get things home when you don't have a car."—Kim Woodbridge, Philadelphia, Pennsylvania

10. Get an app: "Moovit and other transit apps that show when and where the easy transit options will be. I live car free and feel complete freedom because of it."—Sarah Meggison, Tucson, Arizona

Maybe Ditch the Elevator, Too?

We tend to think of elevators as a faster way to move between floors of a building but forget the time we spend waiting for them.

In a 2011 study at Royal University Hospital in Saskatoon, Saskatchewan, doctors reduced their travel time between the various floors of the hospital from twenty-five minutes to ten minutes when they habitually took the stairs instead of the elevator.

Meanwhile, climbing stairs burns more calories per minute than jogging and has half the impact on the joints. A long-term study of ten thousand men showed that those who climbed three to five flights of stairs a day had a 29 percent lower risk of stroke.

Give up the elevator, save some time, and sell your StairMaster!

A Famous Blogger Offers a Word on Where to Stay Put

Leo Babauta blogs at ZenHabits.net about finding simplicity in the daily chaos of our lives. In his words, "It's about clearing the clutter so we can focus on what's important, create something amazing, find happiness." In the summer of 2009, he and his family gave up their car.

A major part of getting around without a car depended on where he and his family finally decided to stay put—where they decided to live. Sadly, so much of the United States is not friendly to car-free living. But places that are friendly to it are often also more fun places to live anyway.

Here is what Leo wrote about his experience:

> *My family (my wife, me, six kids) finally gave up our car. It was a liberating and scary experience.*
>
> *We've been dependent on our automobile for so many years that giving it up was unthinkable. If you own a car, it's probably unthinkable to you too.*
>
> *We drove everywhere: to and from school and work, to music lessons and recitals, to soccer practice and all-day-long games at the soccer field, to family events (which were numerous), to grocery stores and malls and restaurants and*

movie theaters and bookstores and beauty salons (not for me, I'm bald . . . er, shaven), to pay bills and run errands, to go to the beach and the parks. To do anything.

How could we get rid of our car?

Going car-lite

For the last few years, we've been weaning ourselves slowly from the car (actually a van in our case). We went car-lite, gradually, and if you're considering these issues this is what I'd recommend for most families.

First, we sold our second vehicle and learned to make it work with one. At one point my wife quit her job and began homeschooling our kids, which was great because they had their mom home all the time—something most kids don't get. Later I was able to quit my day job and worked from home, reducing our car trips by a lot. Then we moved closer to town, so we could walk and bike more—everything was within walking distance, including the grocery store, beauty salon, post office, beach, movie theater, restaurants, coffee shops and more. Only family and soccer were further away. We used the car very little.

Finally, we moved to San Francisco, and its great public transit was a big factor. We were giving up our car! Note: While many other cities/towns are not as transit-friendly, tons of people have gone car-free in them—walking and cycling and car-sharing are all great options.

Our car-free life

We sold our van (yay!) and didn't buy a vehicle here in San Francisco. A few times we've rented or borrowed a car, and boy, it really reminds me how lucky we are to be without one. It's such a hassle to drive, to find parking, to get a parking ticket (which I've done), to retrieve your car when it's towed (yes, that happened, and yes it was dumb of me), to try to find places when you're driving, pay tolls and pay for parking, to get stuck in rush hour . . . and so on.

We ride buses and trains and walk. We're getting bikes soon, but we decided to do one step at a time. We walk a lot! We purposely picked a home that was a block away from the train stop and has bus lines that are within feet of our front door. We can get anywhere in this city easily.

I often walk aimlessly, just to explore the city. I take Eva and the kids on walks to show them new places that we would never have seen with a car. It's the best way to discover the joys of a new place—cars isolate you and speed you by the best bits.

Buses often have a broad cross section of people riding them . . . [My kids] see so much more of the world than they ever did while isolated in a car. They come shoulder-to-shoulder with humanity in crowded buses, they talk to their neighbors, they smile at people and make others smile.

We are healthier than ever. Walking is amazing. It costs nothing, and yet you get fresh air, see people, see nature, see stores and restaurants and houses and plants you never would have in a car. You get in great shape. My little four-year-old can walk for miles, and sing while doing it. She runs up hills. Granted, sometimes I carry her on my shoulders when she gets tired, but that's good exercise for me. We're also safer than ever—buses are the safest way to travel on American roads.

We spend so much less on transportation. Cars are extremely expensive—not only for the car payments themselves, but for fuel, oil changes, insurance, registration fees, parking costs, tickets, inevitable repairs, the cost of the space to park the car overnight (garages aren't free space), cleaning the car, and health costs (they're unhealthy). When you have so many expenses, you have to work more to pay for those expenses. Cutting them out means I work less, and that's a wonderful thing for me and my family.

I have to give immense credit to my wife, Eva, for being so great during our car-free experiment. Lots of spouses would complain—Eva has embraced and enjoyed the journey. My kids, too, have been great—instead of complaining, they've had fun with me, playing games, singing, exploring, racing. It's been a great journey as a family, and I'm glad we've embarked upon it.

Limitations are actually strengths

People think of giving up their cars, and they immediately think of the reasons they can't—the limitations. But I've come to realize these are actually strengths. Consider.

1. ***Takes longer.*** *Yes, it sometimes takes longer to get places—maybe 20 minutes instead of 10–15, or 45 minutes instead of 25–30. But that's OK, because cars (while faster) are also more stressful. Driving in traffic is stressful. So we go places slower, which is less stressful, more fun. I like a slower life.*

2. ***The weather.*** *Sometimes the weather isn't great—but truthfully, I enjoy getting soaked in the rain. My little ones don't mind either—they love stomping in mud puddles. We are so used to being in our metal-and-glass boxes that we forget how wonderful the rain is. And when the weather is good, cars isolate you from that. You don't get to feel the sun on your shoulders, the wind in your face, the fresh smell of licorice when you pass a certain plant, see the squirrels dart past or the ducks mock you with their quack.*

3. ***Convenience.*** *Sure, buses can be inconvenient—sometimes they're late and you wait and you're late. But think about the inconveniences of cars we often forget: parking, getting stuck in traffic, getting cut off from other people, paying tolls, paying for parking, parking tickets, cars breaking down in the highway,*

car repairs, oil changes, stopping for gas, car insurance, washing the car, the dangers of car accidents (car crashes are the leading killer of American children), the unhealthiness of it for your kids, making a wrong turn and trying to get back on your route, the expense of a car and having to work more just to pay for it, the cost of health care because cars are unhealthier for you and your family and having to work more just to pay for that, just to name a few.

When you look at it like that, considering all the inconveniences of the various forms of transportation, cars don't necessarily come out ahead in convenience.

***4. Groceries.** We walk to the grocery store—it's one block away. We can't carry as much as we can with the car, so we make more frequent trips. That's not a weakness, it's a strength. That means we walk more. Actually, going to the store is uphill, so I sprint uphill. It's a lot of fun and great exercise.*

***5. Doing stuff that's not close.** It's easier to get in the car and go to places, while walking or riding transit takes time and sometimes planning. So yes, you're a bit more limited. I don't see that as bad, once you accept this—it means you do less, which is simpler and less stressful. It means you only go places that are far if they're important. It means you explore ways to have fun near your home. Cars encourage us to take more trips, which pollute more, cause us to be busier, use up more time and money and natural resources. Slowing down and taking fewer trips is better for us, our health, our environment.*

16

HOW TO BECOME A TRUE CITIZEN

The Purpose and the Responsibility of Life

When my daughter, Isabella, was six, we had a little party trick we did for people. I would say, "Bella, what is the purpose of life?"

She would say, "To laugh."

I'd say, "And what is our responsibility?"

She'd say, "To make sure other people can laugh, too."

Admittedly, I had taught her to say that. But the point is, whatever we feel is important in our lives, most lifequesters I meet feel it is their responsibility to help ensure other people can have it, too.

It need not be laughter, of course. It might be food or shelter or freedom of expression or the right to equal protection under the law.

The problem is, not everything we want for ourselves and others can be gotten just by making different life choices. Sometimes the society itself needs to be changed. And to change our society, we may have to become real and involved citizens.

Understanding the Power Social Structures Have Over Our Lives

We have talked about changing our relationships to food and transportation and ownership, among other things. It is great to be able to change your world relationships in your smaller life choices. But some choices you would like to make cannot be made without a change in the society in which you live.

Suppose you would like to choose to breathe clean air that doesn't give your neighborhood kids asthma, but the electricity company burns coal in the nearby power plant. Or suppose you are a person of color and would like to drive your car without being pulled over once as a victim of racial profiling. Suppose you want to buy fresh produce that is actually good for you, but the stores nearby sell only processed food.

What if you suffer the pain of these problems and want to fix them for yourself and others? What if you have overcome the difficulties such problems pose and want to make sure other people can, too?

Putting solar panels on your roof alone will not stop global warming. Working on your own personal bigotry and learning to celebrate difference will not stop institutional racism. Supporting your local farmers' market will not get fresh food into a food desert.

Sometimes we and the people we would like to help are prevented from having the safe, healthy, fulfilling lives we want, not by our life choices but by conditions imposed by the society we live in.

So what can you do?

How to Hack Societal Structures

Here is the thing that stops so many people from feeling that they have power over the social structures that influence the way we live our lives: we think we are separate from those structures, that there is an *us* and there is a *them*. We imagine that the government, for example, exists inside a big black box and that we are outside that box and therefore have no influence on what happens inside it.

Brett Scott, financial activist and author of *The Heretic's Guide to Global Finance: Hacking the Future of Money,* suggests in his book that this us-and-them view of social structures disempowers us and makes it harder for us to change the social structures. Instead, Scott suggests that we think of social structures—government, corporations, institutions—as constructs that exist mostly in people's minds.

If everybody who worked for or participated in a government or a corporation or any other institution got up from the desk, stopped participating, and went home, that particular organization would cease to exist. They really only exist by the agreement of many people to abide by complexes of rules that govern the behavior of the people who participate in them or relate to them. If the participants decide the structures no longer exist, then they kind of don't. (Of course, I am not saying it would be easy to get people to make that decision.)

So a government or society acts a certain way only as long as the people who participate in it agree to follow the rules. While the power to make the rules inappropriately concentrates in some people in these networks (CEOs of big corporations, for example, control vast resources and influence the behavior of many thousands of people), that does not change the fact that each of us is part of the network and at least has some power over that network and the ability to influence other people in it.

When we realize that corporations and governments and other structures are not black boxes, but that we are actually networked into them like computers into the Internet, then we can see we have avenues of influence. If suddenly enough of us decide to stop or threaten to stop obeying the rules of a structure, then the structure is destabilized and must reorganize itself, if it hopes to survive, along the lines of a new set of rules that we will agree to follow.

When groups of us come together to create change, the number of connections through which we influence the network increases

How Thinking "Us" and "Them" Takes Away Our Power

exponentially and becomes equal to the power held by the most powerful of individuals—like CEOs and politicians. Even when our groups are small compared to the size of the entire network, we can disrupt the network enough that it must adapt to our demands in order to continue operating smoothly.

Hence, even though a relatively small number of people—relative to the entire population of the United States—demonstrated in the #BlackLivesMatter protests, the potential for disruption of the entire network was great and so the network has begun, at least in small ways, to adapt. Think of the U.S. Justice Department deciding to investigate the Baltimore Police Department, the increasing use of cameras attached to police uniforms, and the rethinking of the use of guns by some police departments.

How Thinking We Are Part of an "Only Us" Network Gives Us Power

U= Members of so-called US

T= Members of so-called them

In a network, we see ourselves as part of the structures we are seeking to change. All of "us" have connections to and influence over all of "them."

Of course, there are many, many ways to work to change the agreements by which our social structures operate. You can march, lobby your representatives or participate in the political party of your choice, volunteer or work for an advocacy organization, or take up any other of the many options that exercise your power as a citizen.

The point is that when we realize that we, everyone we know, and everyone who shares our concerns are networked parts with influence over the social structures we may hope to change, it is like we have taken Morpheus's red pill in *The Matrix*: we can see our world as it exists and the power we have within it.

To take that power, though, means to be willing to be involved in collective action. It means the well-rounded lifequester may have to accept the mantle of activism and become a truly engaged citizen in order to create the world he wants to live in.

Science That Says Participating as a Citizen Improves the Lifequester's Life

Which way of life is the more desirable—to join with other citizens and share in the state's activity, or to live in it like an alien, absolved from the ties of political society?

—Aristotle, *Politics*

Remember our friend Tim Kasser, author of *The High Price of Materialism,* whom we met way back in chapter 7? Well, back in 2009, Kasser coauthored with a colleague from the University of Göttingen a study on the psychological benefits of being politically active.

They studied people who advocated for some political cause (say, protecting the environment, human rights issues, or preventing wars) via any of a large array of means, ranging from starting a petition to civil disobedience. They studied various elements of activist identity, commitment, and behavior and used statistical analyses to determine how those elements correlated with life satisfaction, negative and positive emotions, meaning and self-realization, and how well people felt they functioned as members of society.

Their studies found that people who were engaged in low-risk political activities were between 1.6 and 3 times more likely to be flourishing than those who weren't. If you think about it, this makes a lot of sense. Being upset and worrying about something in the world without doing anything about it is likely to make you feel bad. On the other hand, feeling as though you are taking action about an issue that concerns you is likely to make you feel good.

Becoming politically active—working with others to change something about society that troubles you—may very well increase your well-being. Here are the internal rewards, Tim wrote in *Yes!*

Magazine, that citizens may obtain from being politically engaged: "a sense of satisfaction, the experience of pleasant emotions and of connection with others, and a feeling of aliveness."

By way of concrete example, here's what two of my Facebook friends wrote about their experience being involved in #BlackLives Matter protests:

> *I was very moved and disturbed and struggled to find a voice and an angle through which I could be helpful. Mostly I have found that listening and bearing witness has felt good. I attended a protest at my school, Hunter, and also marched through the streets in the fall after Eric Garner was killed. We marched into oncoming traffic and stopped the whole street. It was an amazing thing to be a part of. It's hard to know what to do . . . I fear for so many, those I know and love as well as my neighbors and kids I see out in the streets. I guess for now I'm paying attention and trying to stand up for what is in my heart.*
>
> —Ava Dweck, Brooklyn, New York

> *Students at Valparaiso University hosted a "die-in" on campus in the union building. I participated with my young children. We are a biracial, blended family. My blond, pale daughter held completely still for eleven minutes in honor of the victims of police brutality. My olive-skinned baby wanted to nurse, of course, with Mama next to him, and he did. As I fed him and played dead, I prayed for victims of racist police and thought about what happens to the family left behind when every week police brutality claims a life. My biracial brown-eyed three-year-old wiggled and put his head down on the knee of a student who was "dead" beside us. The student stayed completely silent as my son tried to interact with her, and that is what moved me the most—the fact that*

> *the children didn't understand the violent deaths that we were remembering, and yet my son wanted so badly to wake her up and bring her back to life.*
>
> —Sarah Degner Riveros, Valparaiso, Indiana

Using Your Happiness and Sadness to Be an Engaged Citizen

One thing I have heard from many, many lifequesters is that, often, trying to live the Good Life is like trying to swim upstream against the cultural current. You want to get good food, but the entire food system is built for efficiency and profit, not nutrition. You want to share, but the entire economy is built on buying new, not sharing. You want to ride a bike, but the streets don't feel safe.

If you are lucky enough to find your path anyway, then you may want to share the happiness you've found to help change the way things work and make it easier for others to follow in your path. If you are frustrated or saddened by the obstacles, then you can use your frustration to see what needs to be fixed in the culture to solve that frustration for many others.

HOW ONE WOMAN'S SADNESS HELPS HEAL NEW YORK

> *I'm Mary Beth Kelly and I came to be more of an activist for safe streets after the death of my husband, Dr. Carl Henry Nacht. He was killed when the two of us were riding our bicycles on a greenway in New York City in the Hudson River Park in June of 2006.*

So begins a video made by StreetFilms.org about my friend and fellow board member of Transportation Alternatives, an organization that advocates for streets that are safe and good to live on for the people of New York City. Mary Beth is also a founding member of

Families for Safe Streets, a group of families whose loved ones have been killed or severely injured by aggressive or reckless driving and dangerous conditions on New York City's streets.

The narration of Mary Beth's StreetFilms video continues:

> *On a day-to-day basis my husband and I both rode our bikes to our work. All year long he would go between his office and the hospital. If he did home visits, he did it on his bike. It was just a wonderful way to start and finish our days. It meant that we weren't polluting. We weren't even crowding the buses or the subways with our bodies, we were out, you know, on our own.*

Then, on that June day in 2006, the couple was riding home from dinner—*on a bike path, in Hudson River Park*—when a New York City Police Department tow truck turned sharply across the bike lane. The driver of the truck ignored the YIELD TO CYCLISTS sign and hit Dr. Nacht, fatally injuring him. He died four days later.

> *I don't associate being on my bike so much with what happened to my husband as cars and trucks. A tow truck towing a car, you know, that's about this city being kind of owned and operated by the automobile, and that's why I'm fighting back.*

The parks, sidewalks, and streets of New York City, Mary Beth believes, should be places where the people who live there can run, ride, walk dogs, and Rollerblade freely, without having to worry about a vehicle coming up behind them, how fast it's going, or whether it's going to obey the laws and give them the legal right-of-way.

Mary Beth told me over the phone,

I'm a psychotherapist. People come to me because they are unhappy and I work with them on an individual level. We therapists work with why they are unhappy and how to change their thinking about things.

I always knew this, but there is also the larger picture of how we live together in community and in the built environment that supports or diminishes our health and happiness. When Henry died, I realized that no amount of work on self could have stopped that. The streets were contaminated by the domination of the automobile and, as a community, we need to recover from that—big societal problems that affect our lives—too.

Our health and happiness do not just depend on what we do as individuals. They depend on how we come together as citizens to make our laws and rules and mold the built environment. Living the way we want to as individuals depends on our coming together as citizens to help mold the culture.

Mary Beth spends a substantial amount of her time making speeches, writing letters, attending community meetings, and meeting with elected officials. She has managed to redirect her sadness to point a way toward being an engaged citizen whose actions help make the world she wants to live in.

The Families for Safe Streets website says:

We bear witness to our pain and suffering to press for the elimination of fatalities and injuries on our streets. Through our stories and advocacy, we seek cultural and physical changes on our streets and the rapid implementation of Vision Zero [a plan to completely eliminate fatalities due to car crashes]. We envision a city where pedestrians, bicyclists and vehicles safely co-exist, and children and adults can travel freely without risk of harm—where no loss of life in traffic is acceptable.

That is how you use your happiness or sadness as an engaged citizen.

Becoming Engaged

There are so many ways to be an engaged citizen, and like so much else in this book, doing so is a process. Let yourself make mistakes. Let yourself feel like you don't know what you are doing. The feeling of mastery will come. Here are some questions you might ask yourself in order to find your own path to engaged citizenship:

What is the gift that you want to give your community, town, state, or country? How can you make people happy the way you are or want to be happy? How can you help them avoid sadness the way you avoid or want to avoid sadness?

What is the path you can follow to give that gift? Is there a volunteer opportunity? Is there an organization you can find on Google that works in the area you are passionate about? Are there mailing lists you can join that will tell you when people are organizing actions?

Who else cares about what you care about? If you can't find an organization working on your issue or you don't know how to find one, bring together a small group of friends who you think will share your concern. Tell them: *I want to fix this problem in our local/national/world community.* Ask them: *Please, will you help me figure out how?*

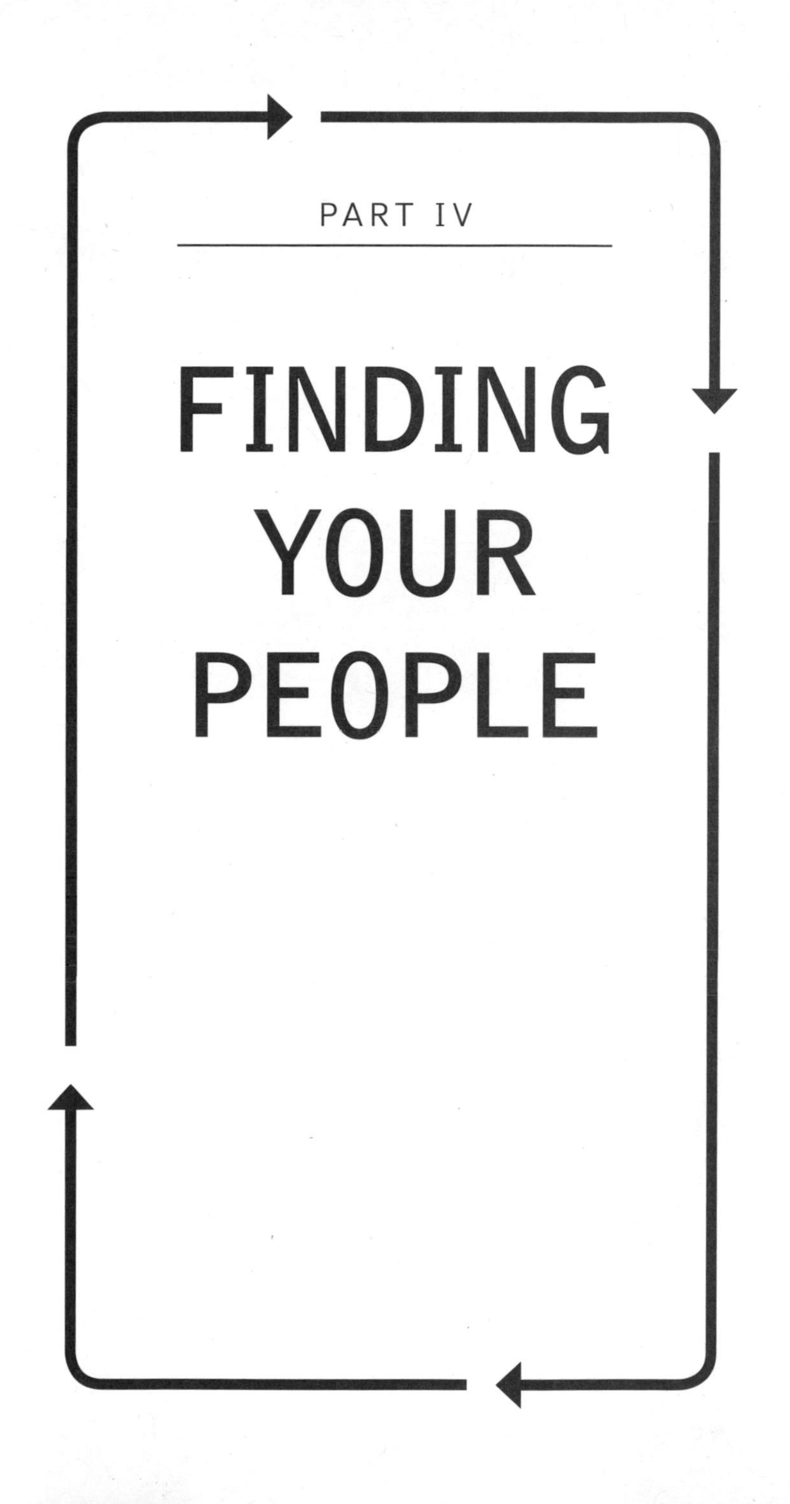

PART IV

FINDING YOUR PEOPLE

17

THE "REAL" THING IN RELATIONSHIPS

As I've mentioned, when you experiment and think about paradigmatic lifestyle change in one area of life, you end up thinking about it in all. When you pull on a ball of string, the whole thing unravels. So lots of people who also think about alternatives to the standard life approaches and non-consumption-based lifestyles—minimalists, simple living folks, DIYers, fixers, pickers—also think about the relative importance of different ways of relating to each other through friendship, romance, and membership in groups.

Our culture and media have very strong prescriptions regarding what we are supposed to want when it comes to relationships with others. What energy we don't dedicate to our careers, the standard life approach goes, should be dedicated to finding the romantic relationship that we will be able to tell everyone is "the real thing." This approach suggests that the most fundamental and important relationship is the lifelong coupling of two romantic partners—the "real" thing.

In this area, as in so many others, some lifequesters are trying to figure out what feels to them like the most authentic way of living in relationship to others. A way of living based on our humanity instead of the societal norms that seem to be leading us off a cliff.

Will a long-term romantic relationship add so much to our personal betterment and our contribution to the world that it should be a nearly exclusive priority in our lives?

The problem is that for many people, a lot of stress is created by the gap between what we are "supposed" to want and what actually makes us happy and is actually available to us. Romantic and sexual relationships come and go, and sometimes the types of relationships we fit into naturally don't seem to fit the prescribed societal mold.

As a case in point, my friend, hero, and sometimes mentor, Julia Butterfly Hill, whom I have mentioned before, once posted on Facebook that she wasn't sure about long-term relationships but she sure would love a great fling. Someone very innocently commented that Julia deserved a great fling but also the "real thing."

Here is the problem with that language: Just as we kind of sanction and pressure each other to have "real cars" and buy "real houses" and have "real careers," and make each other feel bad for choosing nonmaterialistic lifestyles, we seem to pressure each other to have the "real relationship." Just as our culture ascribes value to people according to whether they have a great job, it also ascribes value according to whether they have a relationship that is "real"—which apparently means long-lived, with just one other person, and including sex.

When we don't have "real" relationships, either because we choose not to or because the opportunity is not there right now, we feel ashamed. Women, more fiercely than men, seem to be the particular target of these media and societal norms.

Which leads us back to the big question: Does having a romantic partner amount to our actual value and the contributions we make to our own lives, the lives of others, and the world? Because in truth, some of us dedicate so much energy and time to primary partnerships that we are unable to contribute much to our world and local communities. Doesn't all this suggest, as a culture, that we should redefine normative ideas of the kind of relationship that actually contributes most to the world?

Isn't it possible that we contribute more to our individual and world happiness when we are interconnected with and contribute to larger groups of people, regardless of our romantic status? Isn't it possible that the "real" thing is participating in a more sprawling web of connections—both when we are single and when we are romantically attached?

This is why so many lifequesters appear to be placing a new priority on social interconnectedness—modern forms of what we used to call "community"—rather than exclusively pursuing romantic connection. Because, as we will see, research actually shows that social interconnectedness—as much as and sometimes more than romantic connection—is a predictor of personal success, personal happiness, and contribution to the world.

For these reasons, we are now going to discuss what social interconnectedness, in the Good Life, might look like. Not a "Kumbaya," hokey, by-the-campfire, old-fashioned, back-in-our-claustrophobic-hometown type of social interconnectedness. We are going to look at the modern, edgy, exciting, adventuresome, helps-us-lead-the-life-we-want kind.

We are going to look at why we need social interconnectedness. How it helps us. Why it helps the world as a whole. How to build it. And how it can bring you a happier and more fulfilling life, whether you are preparing for economic collapse, you want a better job, you actually do want to find a boyfriend or girlfriend, or you want to start a social movement. We are going to look at its importance first in the lifequest, and then to everything else.

18

WHEN THE BLACK SHEEP FINDS ITS FLOCK

One Reason Becoming Yourself Is Scary

Listen to this story of lifequester Sasha Beskrowni from Boston, Massachusetts, the challenge of being her Self, and where she found some safety:

> *When I made the transition to veganism, my family was confused, disgruntled, and often discouraging about the changes I was embracing in my diet and lifestyle. I felt like I had to hide that passion of mine from some of the people in my life at times for fear of being made fun of or even pushed to give it up. But I found two amazing friends in particular who supported me, and then the rest of my friends also became a huge source of support for me.*
>
> *I feel like sometimes I can't tell what comes first—my discovery of a supportive friend group or the new phase of life that I find myself in. Maybe I change and grow because of who I surround myself with? Or maybe I seek out supporters?*

Anyway, adding new supporters to my personal community always gives me the strength I need to be able to stick with my older relationships as I change. Despite initial confusion and resulting rough patches that stemmed from that confusion when I turned vegan, my old friends and family remain my biggest constant, and having them means I have had support through so many other phases of my life.

What makes becoming your Self scary?

What makes us scared to live our lives in accord with our prized talents?

What frightens us about demonstrating that we really care about what we really care about?

What is the fear common to thousands and thousands of would-be lifequesters who want to do things like:

- Become a singer.
- Volunteer in a homeless shelter.
- Declare independence from the automobile.
- Become a vegetarian.
- Live a non-heteronormative sexual life.
- Become a writer.
- Or a sculptor.
- Or a painter.
- Embark on a walk across the country just because you want to.
- Let go of money as a primary goal.
- Surf and do nothing else.
- Become an activist.
- Make stuff.
- Hug more often.
- Kiss more often.
- Love more often.

You could say we are scared we won't be able to earn a living or that we won't be supported or that we will fail. And that may be true. But those fears come from something even more fundamental and human about us.

What that is, as we face either the beginning of our lifequest or committing more fully to it, is the understanding that to move forward, we have to become different from who we were. Become different from the person whom the people who love and like us love and like. We know that to move forward on our own individual paths . . .

We have to face the fear that we will no longer be acceptable to those around us, that we won't be loved and liked, and that we won't be supportable.

Sometimes the desire for the approval and affection of others is perceived as a fault. Certainly it can be one—when we are so pathologically scared of not being approved of that we can't live our lives meaningfully or even do what needs to be done. But the fact that embarking on the lifequest is frightening is hardly pathological.

Why we are so scared that no one will like us is because—in spite of wrongheaded misinterpretation of competition of the species and survival of the fittest—we are not fundamentally competitive but cooperative. Thomas Hobbes's idea that in society we engage in a war of "every man against every man" was entirely wrong.

Our whole way of living depends on our working together and cooperating together. Which of us works alone? Gets her food for herself? We may not realize how much we cooperate because we take it for granted. Cooperation is in our biology. Our forebears, the hominids, could not have survived a single night alone.

We also have to find a way to work with needing to be liked, loved, and supported, because that is what makes us human.

Still today, we are defined in relation to the groups we belong to. Back in Part Two, Wanting What You Really Want, we saw that being truly fulfilled does not just mean being authentic but being authentic in membership of a group that values us and depends and needs what is particular to us as individuals.

So the challenge for the lifequester is a balancing act: How do you avoid staying the same in order to gain approval; and, on the other hand, how do you become yourself without cutting all emotional ties and isolating yourself? How do you build a personal community with at least some people who accept, encourage, support, and assist you without necessarily leaving behind everyone who needs your help and courageous example? How do you find a community that is helped by your being who you are?

Stripping Off the Mask

Let's think for a moment of the process of *individuation* or *integration* or *actualization*—different words a humanist psychologist might use in place of *lifequest*. Let's think, in particular, about why Carl Jung might say the actualization process is so scary and challenging in relation to the shifting group of people who make up our personal communities.

As babies and children, we are for so long helpless and dependent on others for care and food and shelter and safety. We cannot walk. Cannot feed ourselves. Cannot protect ourselves. No one depends on the cooperative instinct more than a young human. In order to make sure people want to cooperate in our survival, we quickly learn to do what is more or less expected of us.

Then, as we grow older, our circumstances, talents, and personalities thrust us into additional groups, outside our family—school, college, work, and social networks. The ways we act in these new groups are the result of interaction between what comes naturally to us and what the group needs. We find a role that suits us and we fill it. Again, we do what is more or less expected of us.

Carl Jung called the limited constellation of behaviors—the subsets of our personality, if you will—that we exhibit to meet the expectations of our groups our *personas*. We have more than one; the number depends on the different groups we belong to. For example, you can have one role at work—your work persona—and another role in your family—your family persona. The reason we can't stand being called a family nickname around our colleagues, for example, is because it doesn't fit our work persona.

Knowing we have different personas at different times and places demonstrates that none of our personas embodies the whole of our True Self. We put on and take off our different personas like masks we wear for different occasions. Those parts of our whole Selves that are not useful or acceptable to any of our groups and aren't exhibited in any of our personas are tucked away and hidden in our unconscious. Taken together, those hidden potentials are what Jung called the *shadow*.

Now, what is in our shadows is not necessarily bad or wrong or evil. It is just parts of ourselves that up until now have not been acceptable to our groups or useful to us personally. But because those parts of ourselves are in the shadow—in the darkness and unfamiliar to us and those around us—we can be frightened of them and what they will do to our memberships in our groups if they emerge.

Meanwhile, according to Jung, we have this innate and powerful psychic force inside us, the actualizing force. This is the desire to be who we really are, to integrate all of our capacities. Because of this force, what we have relegated into our shadows must eventually come out. For that to happen, our personas must expand. If the roles we have in our families and our jobs and our friend groups aren't flexible enough to allow that, then our membership in those groups may be in danger.

So there is this tension between our need to become our Selves and the need to play the roles assigned to us within our groups. We are stuck in a push-pull between our need to grow and our need to fit in.

The need to expand can overwhelm the need to fit in, or vice versa, at any time in a person's life, but Jung also identified four types of psychological events that can spur us off on our lifequest (or require us to marshal the huge psychic energy to close ourselves down).

The first of these event types comes when we transition from one life stage to the next: from childhood to puberty to early adulthood to midlife to mature adulthood to late adulthood. The demands of each life stage require us to call on different capacities that we may have previously kept buried in the shadow. We have no choice but to integrate them.

Another event that can spur us to become our whole Selves is a sudden relaxation of the demands of life. Maybe you have the financial security you have always wanted. Or you finally got your degree and no longer go to night classes. Suddenly, you have time. Maybe parts of you that need meaning and adventure come bubbling to the surface and you have no overwhelming responsibilities to distract you from them. You just can't stay on the treadmill, you realize. You have to expand. There must be more to life.

A third turning point occurs when you suddenly realize your moral ideals—the rules you have inherited or chosen for yourself—don't accord with the reality of life. Your background has always told you, for instance, that people of a certain other religion are not to be associated with, but suddenly you are required to interact with members of that religion because of your job and you experience them as good people. This experience causes you to question the rules you used to keep about whom you spent time with and therefore other rules you have taken for granted as well.

The fourth and perhaps most important crisis is when you suddenly discover that something about how you or your group lives causes more harm to others than you can tolerate and you feel ashamed. As examples, you are suddenly overwhelmed by news of white police shooting young black men, or you become aware of the dangers of climate change. You can no longer accept the rules of your groups without question. You can no longer repress the need to find a better way.

The thing is, when you begin to question the personas you live by, either because you've experienced one of these events or for other reasons, it's like pulling at a ball of string. All the contents of your shadow begin to bubble up. You begin to question whether you want to live according to your roles, and because you want to break so many rules, this may make you feel like a terrible person.

This is why the lifequest is scary at first. Because by stripping off our masks, we break our promises to our groups about who we will be. Our first steps often happen under a cloud of guilt, and we have to muster courage to take those steps. We feel like black sheep for breaking the rule of maintaining the status quo.

This is one reason why the heroic act of becoming your True Self is such an act of bravery. It is to venture into the unknown and the vulnerability of demonstrating parts of yourself that your groups have not agreed to accept. Because those parts of yourself have been in your unconscious, you can't know how they will affect your life, how powerful they are, or how much they will disrupt things.

But once they begin to emerge, safety is not far away.

Finding the Safety

Three weeks after a massive, four-hundred-thousand-strong climate demonstration in New York City, my friend Bill McKibben is giving a talk in Cooper Union's Great Hall auditorium. Bill, who helped organize the march, is the founder of 350.org and the author of *The End of Nature* and many other books about climate change and environmentalism.

Bill talks about the successes of the march and how he believes policy makers will take notice. But then he begins to talk about what he thought was another great success of the march. He says, "One of the great satisfactions to me as I looked across the crowds is that what I was seeing was huge groups of new friends gathering together and having fun. They were making their point, but they were having a blast doing it. People who share concerns and passions came out of the woodwork and found each other and supported each other."

Bill is talking about the fact that one of the joys of living by your values can be the almost automatic expansion of your personal community to include other people who support those values. So many lifequesters I've met, when they finally take the risk of embarking on the path of being themselves—whether that means expressing concern about the world by going on a climate march or playing music or volunteering to push for penal reform—have quickly discovered new friends.

Because it turns out that as soon as the lifequester finally embarks, the next part of the story is often that he is finally open to meeting mentors, guides, and supporters. As soon as you finally decide to go to a community garden, as soon as you stand up at work about racism, as soon as you get on your own path, you open up to meeting the other people who are walking it, too.

When the lifequester risks not fitting into her old groups, she quickly finds new groups with which she does fit in.

Three Lifequesters Who Found Their People

Wendy Brown from Dalton, Massachusetts, says: "I start volunteering with Habitat for Humanity. I'm on the job site, on the building committee, doing some design. I start noticing something when I leave these people, any of them, any particular group I'm working with. I'm noticing this pleasant, happy feeling. It's like a kindness vibe that these people have. And I have it, too, so it's like I found my people!"

Megan Strant from Melbourne, Australia, says: "When I went on maternity leave I felt lost and alienated. I was no longer the career woman, but I didn't fit in with the mum crowd. Then I joined a committee that runs a not-for-profit toy library where people pay membership and have access to a huge range of toys to borrow and return. I found it heartwarming and comforting. It became a support network, and I feel so connected to my community rather than like just another stranger. Life feels really different and I am so much happier."

Jere Downs from Louisville, Kentucky, says: "When my son's leukemia treatment ended, I was forty-three and in the worst shape of my life—depressed, and stressed from years of being a 'chemo mom.' I decided to run a marathon to raise money but was not an athlete. I found a community of people, many former non-athletes, and I discovered a lifestyle of healthy eating and exercise. Six years later, I've competed in triathlons and hundred-mile bike rides and raised $15,000 in all for blood cancer research. These athletes are still the dearest community in my life."

We mentioned earlier that the lifequest can be likened to what the mythologist and writer Joseph Campbell called "the hero's journey." The hero's journey itself represents the psychological process of breaking away from the familiar to become who you really are and then, when you are strong enough, returning to the familiar to help others. The reason the story is told over and over in so many forms in every culture is because it represents the deep drive we have to actualize and the process we undergo.

Think of the hero story of Dorothy in *The Wizard of Oz*. She is such a "good" and obedient girl—at least, that is the persona she presents. She gets in trouble, though, when her dog bites Miss Gulch. She tries to stay good, to not let her shadow emerge from behind the mask of her persona.

No one in her family, however, has ever had the courage to stand up to Miss Gulch, even though she tyrannizes the whole county. Auntie Em and Uncle Henry don't stand up to Miss Gulch this time, either. They let Miss Gulch take Dorothy's dog to the sheriff to be put to sleep.

Dorothy has finally met a situation where she can't express her

True Self and personal power while fulfilling her assigned role of "good" girl. She can't hide and conform anymore. Her dog's life is at risk and the stakes are too high. But she is not quite strong enough to publicly exhibit her defiance and stand up to Miss Gulch against her family's wishes. When Toto escapes, Dorothy takes Toto and runs away from home.

What happens at first is exactly what she has always feared will happen, the exact reason she has never come out from behind her masks. Disaster strikes. The moment we venture away from our old ways of being, the tornado blows in. Everything is literally stirred up and torn apart. This is the period where our friend Sasha, above, is arguing with her friends and family about veganism and it feels like no one will love her for who she is.

But very quickly things settle down. The house lands in Oz. Dorothy has already killed the Wicked Witch of the East, which is a good thing. Metaphorically, it means she has killed what had stopped her from being her Self. But now she must face the Wicked Witch of the West—all the scary things that make her want to go back to being the good girl, where others make all the decisions and take care of her.

But now her guides and fellow travelers arrive. First, Glinda the Good Witch magically appears, Dorothy's mentor and guide. Then Dorothy encounters the Scarecrow, the Tin Woodman, and the Cowardly Lion, all of whom, like her, are on quests to live in the world not as hidden half people but as whole Selves. They are companions on her journey—just like the people on Bill McKibben's climate march.

At first, your fellow travelers seem magical and different. They surprise you because they assume you have the capacities you yourself never thought you had, and they support you in using those capacities. They bring them out in you. The Scarecrow, Tin Woodman, and Cowardly Lion treat Dorothy like a leader, not a "good" girl.

So they travel together and protect each other and then find together that they always had what they were told they didn't. They just use it and express it in their own ways. They face and overcome

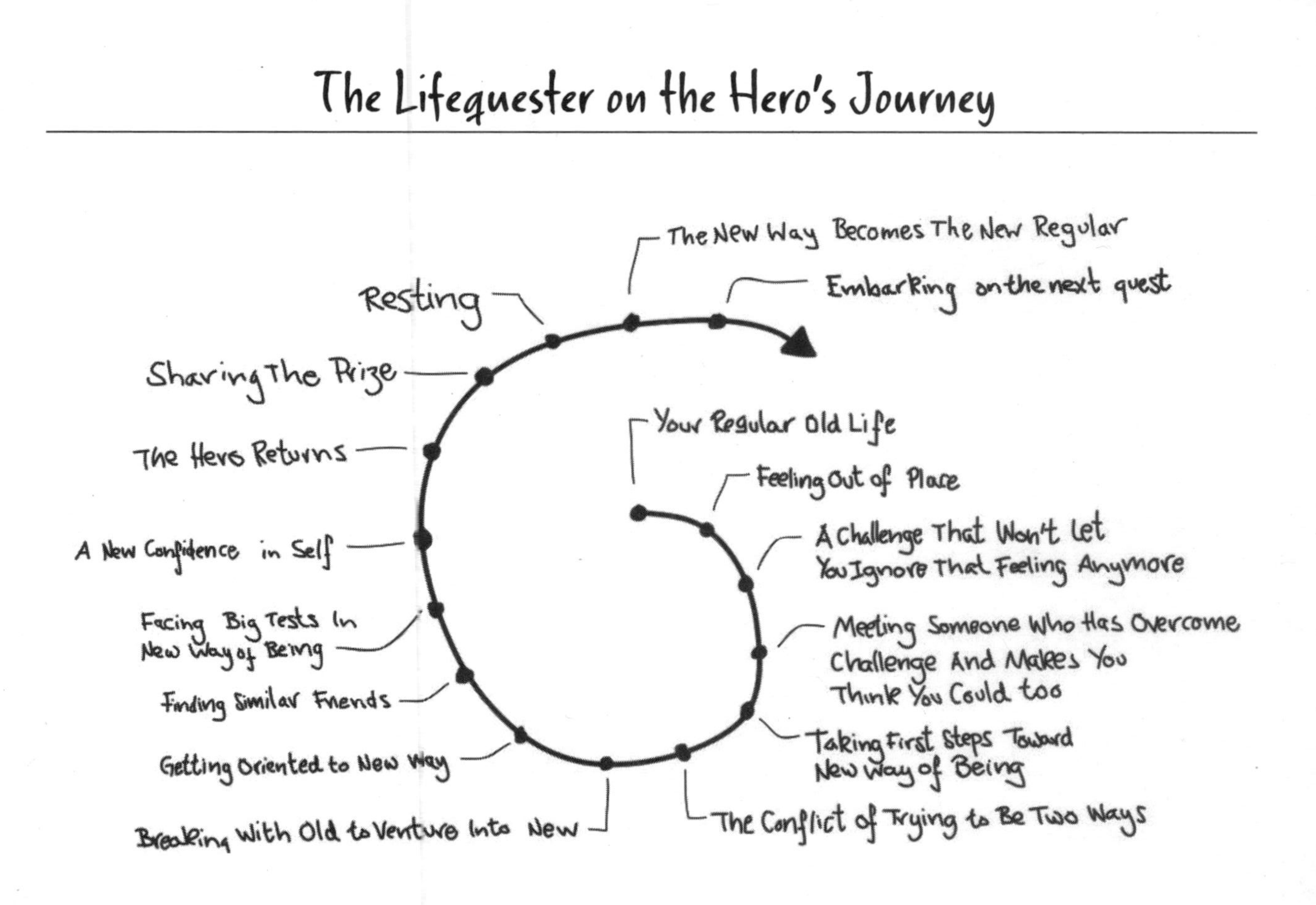
The Lifequester on the Hero's Journey
Your Regular Old Life
Feeling Out of Place
A Challenge That Won't Let
You Ignore That Feeling Anymore
Meeting Someone Who Has Overcome
Challenge And Makes You
Think You Could too
Taking First Steps Toward
New Way of Being
The Conflict of Trying to Be Two Ways
Breaking With Old to Venture Into New
Getting Oriented to New Way
Finding Similar Friends
Facing Big Tests In
New Way of Being
A New Confidence in Self
The Hero Returns
Sharing The Prize
Resting
The New Way Becomes The New Regular
Embarking on the next quest

challenges together. They give each other safety. And ultimately, they kill the Wicked Witch of the West, and Dorothy goes home to Auntie Em and Uncle Henry, strong and proud of who she is.

A Thought Experiment

Imagine that you are supported by close people in all the ways you need to be. That your needs are taken care of. That you can express yourself and help people. Would you even bother with a job? What would you spend your time doing? What would your life actually look like?

If You Become Your Self and No One Is There to Witness It, Did You Really Become Your Self?

Almost every hero story you ever hear is actually the story of some ordinary person breaking away from his personas, venturing into his shadow, and discovering hidden, extraordinary things about himself. And in almost every hero story, guides and companions appear. You don't venture out into a vacuum but into a new group with whom you can try out the parts of yourself you are discovering.

Becoming your Self and expanding your personal community support each other. Acting different attracts new people into your life. Finding new people also causes you to act different.

These are crucially important steps because the challenge is not to be ourselves when we are alone, but to be ourselves when we are in the world. Expanding our "personal community," as a sociologist might call it, gives us the chance to do that. Then, ultimately, we have the strength to bring the whole of our Selves to the rest of our world, where our example encourages and helps people to go on their own quests.

In healthy psychological development we go through this cycle of growing and bringing new parts of ourselves to our worlds over and over again. Our many little lifequests add up to the big lifequest—to fully integrate all of our capacities and potentials and then to use them to help others.

Each time we go through one of these cycles, we add new people to our personal communities and we shed some of the old members who don't mesh with our whole Selves. Because the people in our communities reflect who we really are, we continue to add, shed, and keep people as we move toward realizing our authentic Selves. The closer we get, the less leaving and rebuilding of our communities happens.

Accepting this process is part of what makes the lifequest heroic. But there is a flip side to it as well: we bring to the people who remain parts of our communities something that is crucially useful. Like Dorothy's ability to stand up to Miss Gulch. When people like Sasha have the courage to expose their values to their family and old friends, their family and friends have the opportunity to discover the courage to live by their own values, too.

Finding our people is both a central benefit and a precondition of the lifequest. It turns out that:

> The unexpected reward of the lifequest is connection to new groups of people who share and celebrate your values.
>
> The unexpected result of connection to groups of people who share and celebrate your values is support on your lifequest.
>
> When you embark on your lifequest and join with others, you automatically form a group that supports others, too.
>
> Most important, there really is no lifequest without community, because the lifequest is about becoming yourself in relation to the world.

Find Your People

Where do the people who are already like the person you want to be hang out? Go there. Hang out. Go regularly. Until you meet someone.

19

SUCCESS IN EVERYTHING THAT MATTERS

A Unifying Theory of Selfishness and Altruism

Suppose building personal community as a means of fulfilling a lifequest doesn't interest you. Suppose you don't care about the lifequest, per se. Suppose you just want to:

Make more money . . . Start a social movement . . . Get a better job . . . Make your neighborhood safer for elderly people . . . Date more people . . . Build more support for local youth . . . Have people to celebrate your successes with . . . Prepare for economic Armageddon . . . Build a social safety net . . . Feel less lonely . . . Help your children be more successful . . . Always have something to do on Friday night . . . Have endless couches to sleep on if it all goes south . . . Feel purposeful regardless of employment status . . . Do something that helps others as much as it helps you.

Because it turns out that—lifequest or not—building your personal community is one of the biggest predictors of life satisfaction and success in almost everything that matters.

Yet, for so many of us, robust personal community is a casualty of the standard life approach. We move away from the people we love. We work too many hours. Bit by bit, our old connections fray and we don't have time to build new ones.

On the other hand, so many lifequesters I've met have rediscovered personal *inter*connection—groups of in-person (as opposed to online) friends who are also friends with each other. In fact, building a robust *interconnected* personal community often becomes *the* primary goal of lifequesters.

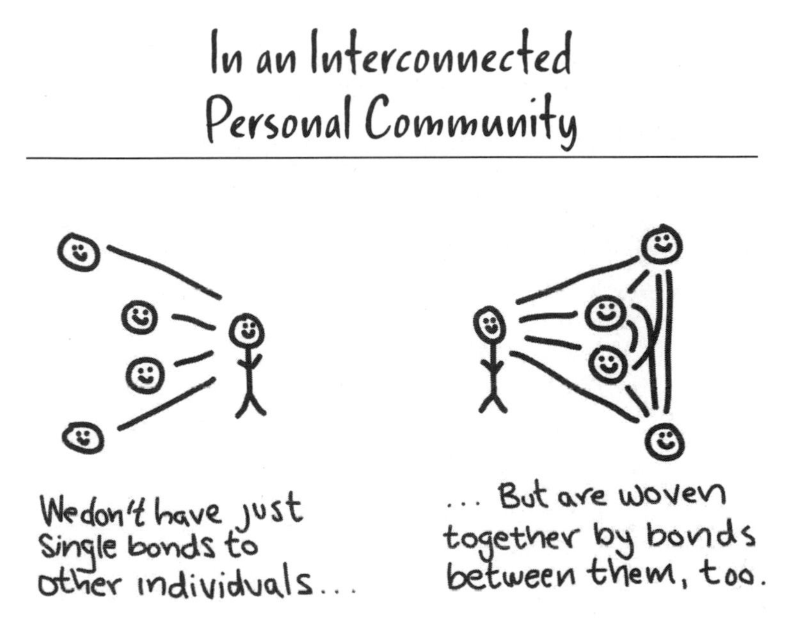

We don't want to admit it, but so many of us are terribly lonely (one in three Americans over forty-five report being chronically lonely), and building personal community alleviates that (actually, in ways that romantic partnerships won't—more on this later). We will see later how that resulting community can end up doing something amazing for the world.

So:

What if the key to success in life comes not when you get the right job or the right romantic partnership but when you build the right personal community? What if our personal community, as opposed to a job, is the place where our skills and talents can be best used and most appreciated?

Meanwhile, selfishly building strong, interconnected, in-person communities to get those benefits for yourself automatically brings those benefits to the people you include, too—making it especially satisfying for the lifequester.

Later, we will discuss the science that proves that the benefits of building personal community ripple out well beyond the people we are personally connected to and into the rest of the world. This happens because when you form a social bond with another person, not only do you and the other person benefit, but so do the people whom you and the other person are already connected to.

Therefore, building and maintaining a network of strong, interconnected relationships helps you, everyone in your network, everyone in their networks, and—actually—the whole world to have better lives, be more effective and happy, and get the things they want.

Building personal community is an amazing act of both self-preservation and generosity at the same time. This is why it is the unifying theory of selfishness and altruism.

The Two Secret Ingredients of the Most Effective Personal Community

1. REGULAR, IN-PERSON CONTACT WITH SIX TO TWELVE FRIENDS

Most of us are pretty proud of our 1,306 Facebook friends. For sure, Facebook can be fun and worthwhile for lots of us. But most of us need more strong, *in-person* bonds.

The problem is that most Facebook friends provide what sociologists call *weak ties*. Don't get me wrong, weak ties are helpful in their own way. They are good for information exchange—like finding out about a job you wouldn't otherwise know about. Weak ties bring new information, while the people with whom we have strong ties tend to know a lot of the same things as we do. They don't bring much new information.

Going back to our weak ties for a moment, studies show that Facebook friends *can* contribute to your *sense of* well-being and your *sense of* social support. The question is whether the increased *sense* of social support reflects increased *actual* social support. Will your Facebook friends show up when the shit hits the fan in your life?

Even two thousand faraway Facebook friends aren't as good as one strong, in-person friend when you need someone to sit in the emergency room with you or when you need someone to celebrate a life success with.

This is why, in general, loneliness researchers have determined that the overwhelming proportion of weak ties we tend to form on Facebook don't take away feelings of isolation that plague so many Americans and Europeans. Between 1985 and 2004, the proportion of Americans who said they had no one with whom they discussed important matters tripled to 25 percent. Another 19 percent had only one person to confide in, often their romantic partner.

Added together, that's moving toward 50 percent of our country having a maximum of one really close friend. Europeans, the research shows, are catching up in the loneliness stakes.

What counts as a strong tie? One good method sociologists use to figure that out is to ask, for example, how many friends you feel you could turn to in order to discuss a romantic heartbreak, borrow money, ask for help moving house, or talk to if you needed cheering up.

How many should you have? One study showed that the happiest college students are in regular contact with between six and twelve close friends. Students with less than that number tended toward loneliness. Students with more began to feel overwhelmed with their social obligations.

Most loneliness researchers agree that the number that average citizens of the northern hemisphere would be able to count is way too small. That's why, in order for us to be more happy, effective, secure, and helpful, most of our personal networks need to include stronger in-person relationships.

2. REGULAR, IN-PERSON CONTACT BETWEEN YOUR FRIENDS

Suppose you *are* able to jot down a healthy list of people who would loan you money, help you move, or support you through a breakup. That would indeed make you socially connected. That's a good thing.

But there's a difference between having a lot of friends—being connected—and having a personal *community*—being *inter*connected—and it makes a big difference to well-being. The clincher, when it comes to being socially *inter*connected, is how many people on your list would turn to *each other* for help. How many of the people in your support network also support each other?

Studies of college students show that the more interconnected a student's friend network, the less likely the student is to be depressed. Other research on teenage girls showed that the more "clustered"—interconnected—their friends were, the less likely the girls were to contemplate suicide. In general, studies find that people whose friends know each other are less likely to experience feelings of loneliness.

Why Does Interconnectedness Matter?

According to sociologist Robert S. Weiss of the University of Massachusetts, who founded the study of loneliness, we have two distinct needs for companionship. *Emotional loneliness* is the need to be deeply connected to or understood by a romantic partner or best friend. But Weiss also proved that we experience *social loneliness,* the biological and psychological need to be part of an interconnected network of real relationships.

We need to be bonded to people who are also bonded to each other. Weiss showed that social integration in a personal community and spending alone time with unconnected best friends or a partner can't replace each other.

Think of it this way: A child rejected by peers can't find consolation in spending more time with Mom. Nor can the absence of Mom be helped by the attention of playmates. Children need friends to play with and mothers to care for them. Weiss showed that, similarly, adults need a social network to feel engaged and meaningful and best friends or romantic partners to feel safe and secure.

The two types of loneliness—emotional and social—even feel different. In both, we experience mild depressive feelings and general dissatisfaction. But emotional loneliness, according to Weiss, includes anxiety and apprehension. Social loneliness includes restlessness, boredom, and feelings of exclusion.

Think about it. If you had ten unconnected close friends or partners, you could endlessly hash out your life problems over coffee, telling the same stories repeatedly. If you had ten interconnected friends or partners, you could have just a couple of coffee dates and a nice dinner party with the whole group where everyone shared about their lives but then there was plenty of time to joke around and gossip.

The Science of How Personal Interconnection Makes Your Life Rock

Here is a sampling of the science that proves that building interconnectedness substantially helps our own lives:

Interconnected people are safer: According to Juliet Schor, a Boston College sociology professor and the author of *Plenitude: The New Economics of True Wealth,* interconnected people were more likely to survive and fare well during the 1995 Chicago heat wave and Hurricane Katrina.

Interconnected people sleep better: A 2011 University of Chicago study showed that lonely people rouse more often during the night, possibly because early humans who were alone evolved to be more vigilant against potential danger in the night.

Interconnected people act smarter: A 2009 paper in the journal *Trends in Cognitive Sciences* showed that because lonely people devote more brain power to guarding against social threats, they can give less attention to mental tasks and therefore exhibit lower cognitive ability.

Interconnected people live longer: A 2012 paper in the *Archives of Internal Medicine* reported that people over sixty who reported feeling lonely, isolated, or left out were 45 percent more likely to die over the course of a six-year study.

These benefits result from the fact that having a personal community makes us feel safer and more purposeful and also allows us to benefit more from our own acts of generosity.

We Feel Safe:

Interconnectedness makes us feel safer, which in turn helps our bodies and brains function better. If you have an interconnected group of friends, chances are, no matter if you lose your job or get kicked out of your house, you will be taken care of.

Knowing this makes you feel less vulnerable to danger. Levels of stress hormones like cortisol and adrenaline, which can undermine your health in the long term, go down. Your brain can pay attention to what you're doing instead of worrying so much. You sleep better. You think better. You live longer.

We Feel Purposeful:

Being part of a personal community gives you a venue for using your personal gifts, whether you have a meaningful job or not. If you are an organizer, you have someone to organize. A listener has someone to listen to. A singer has someone to sing to. Your sense of meaning and life satisfaction goes up.

We Benefit from Our Own Generosity:

Being interconnected also means you are more likely to benefit from your acts of selflessness and generosity. If you do a good turn to Joe—within the context of a group—then Harry, Charlie, and Sally are more likely to witness it and tell the other members of the group that you are a good actor. As a result, people will be more likely to do you good turns.

This may be one of the reasons why people who are more interconnected are more likely to be employed and earn more, according to John Cacioppo, a professor at the University of Chicago who cowrote *Loneliness: Human Nature and the Need for Social Connection*. He writes: "What is the key to health, wealth, and happiness? . . . You are fundamentally a social being. The key to it all is to form strong social ties that are meaningful and satisfying, both to you and those around you, near and far."

Even Cooler Science of How Your Interconnection Makes Everybody Else's Life Rock

The wonderful news is that when we build interconnection in our own lives, we automatically build it in the lives of many others. If you have ten friends and a new friend has ten friends, the single bond between you and the new friend has substantially increased the chances of your network creating one hundred more bonds.

Here's some scientific proof, taken from statistical studies cited in Robert Putnam's book *Bowling Alone: The Collapse and Revival of American Community,* that interconnection helps the world:

> **Interconnection helps children thrive:** Statistically, adult interconnectedness in a locality is a better predictor of kids' scholastic achievement than material wealth, the education level of the adults, race, or religion.
>
> **Interconnection reduces crime:** Measures of interconnectedness proved more accurate in predicting regional murder rates between 1980 and 1995 than educational level, number of single-parent households, or income inequality.
>
> **Interconnection increases economic prosperity:** A study of high-tech companies concluded that growth in California's Silicon Valley outstripped growth in the Route 128 corridor outside Boston because the Silicon Valley entrepreneurs were more interconnected.
>
> **Interconnection improves community health and well-being:** Studies by sociologist James House conclude that the positive contribution to an area's health made by social interconnectedness is comparable to the negative contributions made by smoking, high blood pressure, and obesity.

Just as there are a couple of basic principles for why interconnectedness helps individuals, so there are a couple of basic principles for why interconnectedness helps the world. Those principles are trust, generalized reciprocity, co-regulation, and social capital.

Familiarity Breeds Trust:

Whenever we spend time with people and support them and let them support us and loan them five bucks and get it back, our interactions start to become smoother and easier. There is a lower "transaction cost," meaning the time and effort required to investigate whether the other person is trustworthy before we relate to them. With familiar people, we have already established the trust.

Trust Breeds Generalized Reciprocity:

If we are interconnected, then we also trust people whom we haven't yet interacted with. *A friend of a friend is my friend* means that your trustworthiness has been certified for me by someone else, even if I don't know you. The transaction cost of friends of friends is lowered, too. This is the phenomenon that I mentioned earlier called *generalized reciprocity* by sociologists.

Once we have trust and a sense of generalized reciprocity within a group, we find that such trust begins to extend beyond the periphery of our group. We feel safe in our group and know they have our back, so we can risk being kinder to strangers. We can tip more because we feel less insecure. We can help an elderly person carry groceries home because we know someone else will help us if we are too short of time to get our own errands run.

Generalized Reciprocity Spreads Infectiously:

What's even more amazing is that, because of the principle of *co-regulation,* our feeling and demeanor of trust in the process of generalized reciprocity spreads like magic. To understand the fascinating principle of co-regulation, let's first discuss self-regulation.

Self-regulation means changing our behaviors as a result of stimuli within us. We eat when we have the internal feeling of hunger. We sleep when we internally experience tiredness. The point is that it has nothing to do with anyone else. It is self-regulation. I don't go to the bathroom because your bladder is full. I go because mine is.

It turns out that social species like humans also regulate one another. I respond and behave differently depending on *your* internal state. This actually happens on a physiological level. So, for example, when female chimpanzees are sexually receptive, the testosterone levels in nearby male chimpanzees go up. By grooming each other, chimps promote harmony in their entire troop. The chimps' behavior changes according to the internal state of other chimps.

This principle of co-regulation also applies to trust between humans.

If you and I are part of a loosely knit group of friends and we play guitar regularly or have dinner parties, and we learn to trust and do right by each other, we begin to generalize that trust to strangers. We've established that. But the fascinating thing is that sociologists have established that those strangers then begin to treat other people with the same trust and kindness, because of the principle of co-regulation.

A stranger begins to act trustingly because of my internal state of trust. If I am pleasant to the cabdriver as I slide out of her taxi, she is pleasant to the next passenger who slides in. If I shout at the cabdriver, she shouts at the next passenger. If I help an elderly person with his groceries, he might offer to watch the neighbors' kid while they, in turn, run to the grocery store. It's not a rule cast in stone but a tendency. Co-regulation.

Generalized Reciprocity Becomes Social Capital:

Imagine that you have or are building a network of interconnections and that the people within it trust each other and then begin to generalize that trust. Think of the thousands of trusting interactions that your group of friends has each day. Think of how they co-regulate the strangers they meet. The trust builds and builds.

Trust then becomes a sort of *social capital.* It is like stored money—financial capital—that can be lent. This means that instead of having to overcome suspicion for each other, we will come together to build a business or a social movement to change the world. We will knock on the doors of our neighbors if we don't see them. We will share our children. There are more eyes on more situations.

More people willing to come together to solve common problems. More children watched out for. More elderly people assisted. More of all the good stuff. And *you can help build this by just making some friends and helping them spend time together.*

Why Being Cooperative Makes Humans Competitive

Most of us have been told all our lives that competitiveness is what drives our society toward greatness. That it's in our natures to go it alone. That the need for connection is really a weakness. That it's human nature to be kind of brutish, so we will get hurt if we aren't brutish, too. That everyone is out for themselves. Survival of the fittest and all that.

But if it is truly human nature to be selfish and act alone—the philosophical basis of so many of our standard life approaches—why does the science show that interconnectedness and cooperation make us happiest and most successful?

Evolutionary psychologists say it is because, contrary to prevailing philosophy, we are not actually biologically coded to compete. If we were and human nature was entirely individualistic, our ancestors would never have survived the dangers of the rain forest.

Our survival as a species in the face of much more dangerous competitors, like saber-toothed tigers, depended on the fact that we connect and rely on each other. We did not evolve to be in constant competition with one another. We evolved to be connected and to cooperate.

COOPERATION AS A SURVIVAL STRATEGY

Think of the life of a great white shark, to whom cooperation is not important. She swims the ocean looking for tasty seals. She has three hundred serrated teeth. She can scent the blood of an injured animal from miles away. She can swim twenty-five miles per hour to overtake her prey. She has amazing adaptations for the solo hunt.

Now think of chimpanzees. They don't have three hundred teeth, can't move with speed for very long, and can't detect their prey from great distances. Evolution, in other words, has given the chimp few of the tools she needs to hunt and live alone.

What she can do is cooperate. She can strategize and work with other chimps to surround and corner prey. She then fairly shares food with others, because if cooperation is how you survive, you need to keep your fellow chimps nourished enough to be able to cooperate with you. The chimp species' survival is directly related to its social skill and ability to form reciprocal bonds.

Now think of our human ancestors when they lived in the wild. Humans have dull teeth, lack claws, and aren't particularly strong. Nor do they have the chimpanzee's long limbs and ability to climb up a tree and hide. Even more so than chimps, early humans had literally nothing to protect them from a tiger—except each other. They could not go it alone.

This is the reason, evolutionary biologists postulate, that the human brain evolved to be so large. Ninety percent of our higher cortical functions are dedicated to the ability to relate to other humans—to language, cooperation, and interpersonal dynamics. It was the need to effectively cooperate with other people that made us so smart.

HOW FEELING LONELY KEPT EARLY HUMANS ALIVE

Just as we experience hunger when we get no food, thirstiness when we get no water, or horniness when we get no sex, we experience the sensations of loneliness if we are detached from our social connections. This is because, the evolutionary psychologists believe, human survival depended so much on interconnectedness that we needed the strong, unpleasant feeling to ensure that we stayed with our groups.

The idea is that when, say, we capture or find a tasty feast, we might be tempted to hog the source of food all for ourselves. But loneliness drives us back to the group and forces us to share our catch. That's because even though it might be better in the short term to hog all the food, in the long term, it's good to have strong bonds with the group so that they will help us fight off dangers, share their food with us, help us when we are sick or injured, and work with us to bring up our young.

Now, you may think that this drive is archaic and anachronistic. That there are no tigers in the woods and no bush meat we have to hunt for, so the drive to be part of a group is no longer necessary, a vestigial emotional need that exists in our bodies but no longer serves a function, not unlike the appendix. Being driven back to the group to cooperate is not what's most important in modern society.

As we have seen, the science shows the opposite—we fare best when we have that strong, interconnected personal community. The more valuable we are to our community, the more it rewards us and we thrive. In fact, this is another reason why developing our unique talents and passions—becoming our Selves—and using those potentialities in service helps us thrive.

If we are competitive, it is not to show that we are the smartest or the strongest. We are competing to show that we are the most useful to our personal community. That we have the most to give to others. That we are most willing and able to be of service.

LONELY PEOPLE MAKE PEOPLE LONELY

If everything we are saying is true and we are all so lonely and we really want bustling personal communities, how come we aren't building them?

Well, imagine again our evolutionary ancestors living in the jungle. Imagine one of them has become detached from the group and feels lonely. That triggers feelings of danger and insecurity. Those are feelings that many of us who don't have robust personal communities feel perpetually.

Now suppose that our lonely ancestor encounters a stranger in the jungle. You'd think that if you were lonely you would be glad to make a new friend. But metaphorically speaking, if you meet a stranger in the jungle on a dark night and you are alone, the smart thing may well be to be cautious and maybe even try to appear a little frightening.

Biologically speaking, that is the reaction loneliness triggers. Part of its biological function is to remind us that we are detached from our groups and that no one has our backs. When we are with our group and safe, we can afford to be welcoming and generous. When we are not with our group, loneliness predisposes our brains and bodies to believe that any stranger we encounter is a potential threat.

Zap back to the twenty-first century, when many of us belong to no groups and verge on chronic loneliness. It turns out that our loneliness is biologically forcing us to be shy, anxious, suspicious, and hostile. Since everyone is a stranger, everyone feels like a threat.

This is part of what causes us to stand at the edge of the party without joining in and having a conversation. Or to sit next to someone on the subway and feel nervous about talking with them. Or worse: to treat a person who approaches in a friendly way as though he is unfriendly and scare him off.

We don't know each other because we are scared of each other because we don't know each other. To add to the problem, the principle of co-regulation means that we pick up on each other's fear. I'm wary of you, so you're wary of me, so I'm wary of you, and we all drink alone. Uninterrupted, the loneliness cycle gets worse and worse. The good news is that we can break it.

Step 1: Lower Your Guard—Expect the Best

The less connected we are, the worse our expectations of other people are. The problem is, because of the principle of co-regulation, people tend to meet our expectations exactly.

For this reason, a University of Chicago analysis of all the methods developed by psychologists to combat loneliness found that "cognitive restructuring"—the changing of our thoughts—was the most effective. Changing the way we think of other people turns out to be more effective than learning social skills or even increasing opportunities to meet new people. People who expect the best in social situations get the best.

Since we get what we expect, we should expect what we want. One way to do this, according to a study at the University of Southampton in the United Kingdom, is to be particularly and deliberately nostalgic for old friends and grateful for present ones. Deliberate nostalgic recall of social ties reduces feelings of loneliness, helps us have higher hopes for new relationships, and makes us feel less defensive and better able to make friends.

To make this principle work for you:

> Write down some events like birthday parties or particularly great dinners or times when you felt surrounded by love and companionship.
>
> Write down friends whom you're grateful for and who make you feel safe and protected. Write down the way you feel when you are having your best times with them.
>
> Summarize the list on a card you can carry.
>
> Whenever you are among strangers, look at your card. Replay your nostalgic memories in your mind. This tricks your brain into not being lonely and causes you to lower your biological guard. You can approach people with the highest of expectations.

Remember, the research actually shows it works. Think about it. It is much harder to approach a stranger and start a conversation if you are alone than if you are with friends. Having your best relationships in your mind gives you social confidence and makes you seem less awkward, more open, and more attractive.

Step 2: Lower Everyone Else's Guard—Offer *Your* Best

Just as loneliness increases suspicion in us, so it does in everyone around us. If we want to build interconnectedness, the loneliness experts tell us that we need not just to lower our own guard but to help everyone else lower theirs. All that takes is making their brains believe that we are making them feel good and safe—the way a friend would.

Demonstrate how others find you safe: Conjure warm feelings you have shared with other people in similar situations. For example, if you're talking to a stranger on the subway, mention a pleasurable experience someone else had with you on the subway: "You know, I had a conversation with this other guy on this same train just a week ago . . ."

Confide a weakness: Don't talk about how in debt you are or how you just got fired—those will trigger warning bells—but confide a personal shortcoming that will make you seem less intimidating. "You know, I'm so forgetful . . ."

Flatter genuinely and celebrate victories: Express real enthusiasm for people's passions and successes. This reminds them that they are strong and safe and not in danger from strangers.

Elevate their mood by triggering happy memories: "Who makes you feel the warmest?" "What do you like most about your partner/parent/best friend?" Stimulating their nostalgia will lower their loneliness guard, just as it does yours.

Offer to make introductions: Demonstrate how valuable you think they are by flattering them to others.

20

BUILD IT AS YOU LIKE IT

You Don't Have to Join Your Grandma's Knitting Circle

As we've seen, a robust, dynamic personal community can support and encourage our lifequests and help bring success in virtually everything else that matters, too. Which leaves just two questions:

1. How do you actually build or improve community connection in your life?
2. How do you make sure you have a community that actually supports your growth instead of stunting it?

The problem is, for some of us, the prospect of membership in "communities" or "groups" sounds horrifying. When we don't fit the sometimes rigid norms, we can feel cramped, excluded, marginalized, or even abused in "extended families" and "neighborhood communities." For some, the most exciting change brought by modernity is the ability to leave, and life's happiest and defining moment was packing a suitcase.

I don't mean to trash nuclear families and neighborhood communities. If you dig those structures, can find them, and can be the person you want to be within them, they may well provide the social interconnectedness you need.

For others of us, the good news is that gaining the benefits of social interconnectedness does not mean we have to return to some Grandma-sees-you-every-weekend-and-nags-you-about-marriage model of community. We aren't talking about living in your hometown where people gossip about how you are different. Nor are we necessarily talking about joining bowling leagues, knitting circles, or reading groups (those are good sources of interconnectedness, too, but only if that's your thing).

Even Robert Putnam, the Harvard professor whose book *Bowling Alone* most famously bemoaned the "collapse of community," wrote that we need not be nostalgic for traditional institutions and forums for interconnectedness. Instead, he said, "We desperately need an era of civic inventiveness" to create modes of interconnectedness that suit the way we have come to live and the people we have come to be.

That inventiveness is developing, and new forms of interconnectedness have begun to arise. Sociologist Barry Wellman, for example, at the University of Toronto, writes about "networked individualism"—interconnection without membership in either big clubs or small family groups. Liz Spencer and the late Ray Pahl at the University of Essex studied "personal communities" comprised of people "chosen" from subjects' family and friends. Writer Ethan Watters coined the term "urban tribes" to describe the interconnectedness of twenty-, thirty-, and forty-year-old never-married city dwellers.

The point is that we can be interconnected in whatever way suits us and makes us happy. Gaining the benefits of interconnection simply requires us to be a part of a web of people whom we trust, who trust us, and who trust each other.

Being interconnected need not mean being in a nuclear family or an urban tribe. It can mean either or neither or even both. When it comes to the benefits to the individuals and the culture, it is the interconnectedness that is important, not the form that it comes in. The water is more important than its vessel.

Envision Your Community As You Would Your Career

We all spend so much time envisioning what our career will look like—imagine if we gave that same aspirational energy to the kind of social connection we would like. Who are you? What don't you have in your life that would help you grow, be happy, and give to the world? What kind of group might support and bring out what is hidden within you? What would be fun?

That isn't to say that your personal community must always be composed in the same way. What our personal community looks like should change according to what we need from it. But what emphasis do you want your personal community to have? Here is a sampling of types other people have built:

Hangout Crews:

The term is mine, but the type is based on clusters discussed in Ethan Watters's book *Urban Tribes: A Generation Redefines Friendship, Family, and Commitment*. Mostly, the people get together for dinner parties or hikes or other activities; hang together; drop into each other's houses; and provide one another with social, emotional, and practical support. To make a group like this, Ethan told me, "The key thing is ritual. Tuesday-night dinners, say. Maybe an annual trip. Do the same thing over and over again and it lends itself to the building of a story about the group that people value."

Change-the-World Groups:

My favorite example of this type involved my friend May Boeve and the other Middlebury graduates who started the climate change organization 350.org with Bill McKibben. May and a group of other students who were concerned about climate change regularly had dinner together on Sunday nights.

They bonded over their values and the issues they cared about and became best friends. When college was over, they had built so much trust, affection, and shared vision that it was an easy leap to begin the climate change organization that eventually became 350.org. "I wanted to expand the circle so other worried people had a place to go," May told me.

Chosen Families:

A law professor at the University of Kentucky, Melynda Price, turned thirty-five and didn't want to wait until she found the right partner to have a baby. She was far from her given family in New Orleans and new to Lexington, so having a personal community willing to behave like committed family members has been crucial to her.

How did she build her group? "I decided to never stay home alone. I always went to the same coffee shop and worked. People got to know me. When the invitations started, I never turned one down. I always answer the phone and always open the door. I try to always be there. I will pick up the slack for other people if they need me to."

In short, Melynda was to other people what she wanted them to be to her.

Of course, these are in no way the only manners in which you can make your own community. But to build the one that suits you best, begin by understanding what you want.

Do This!

Create a Vision for Your Tribe

Write down what you want your community to be. Take some time to brainstorm. What role in your life do you want your community to fill? Security? Excitement? Challenge?

What days or nights do you want to be social? Do you want your group to be mostly reliable or mostly crazy and fun? How many really close friends? How many pals to go out with? Who would you like hanging out at your house for Sunday brunch?

Maybe you have lots of confidants but now need someone to party with. Maybe you have lots of people to party with but no confidants. Maybe you want support. Maybe you want to talk about the world's problems more. Maybe you want to play badminton.

Take some time over a few days to begin to jot down a picture of what you would like.

More Personal Communities

A HOUSE-FIXING GROUP

We are involved in a house-fixing group that grew out of our co-op grocery store. It's basically guys (mostly) that go around and fix each other's houses. There's an unwritten/unlogged "time bank" sort of system, but really the goal is to learn from each other how to DIY your house and drink a few beers, talk, and just have fun. Also, you can borrow each other's truck or whatever if you only have a tiny car (like we do). The highest "tech" that we employ is a Google spreadsheet of what car and tools each person has so if you catch yourself saying "if only I had a miter saw . . ." or whatever, you can remember who has one.

—Megan Squire, Elon, North Carolina

A SPIRITUAL GROWTH GROUP

I have been getting together with a group of friends on and off for several years now. We usually meet on Sundays depending on schedules. We're flexible. Sometimes we meet in the park. We call it our Sacred Circle because we try to keep the talk to what we call our "spiritual" growth. No one leads. We all contribute. We do it instead of going to a church service where there is a teacher. We practice deep listening. It's a powerful way to be together as friends, share our lives, and at the same time to grow and have a sense of family.

—Perch Ducote, New York, New York

A MADE FAMILY

As a Single Mother by Choice (SMBC), I set out to create an extended family for my newborn. All of my immediate family (which is a very small nuclear unit) lives out of state. My son has two godmothers and their husbands, two godsisters (one older and one younger), a godfather and his wife, what I call a wisdom mother. In addition, we are part of a community garden where he can approach fellow gardeners and develop his own relationships.

—Keely Nelson, Mount Kisco, New York

Understand a Strong Personal Community's Building Blocks

Once you have an idea of what you want your interconnectedness to feel and look like, you can add to the vision some of the elements that science says it needs to make it structurally strong.

Two researchers at the University of Reading took a look at what they called people's "personal communities," the constellation of important relationships surrounding their subjects. They analyzed the elements of the personal communities and also measured the well-being of their subjects.

Here is what they, and other researchers, have determined are the robust elements that must be included in your personal community in order to make it effective in your life:

> Don't rely too strongly on professionals like therapists for your personal support. Build up some soul mates and confidants as well.
>
> Don't let your community be comprised of only friends and family of your romantic partner. It is not as rich as you each having your own but overlapping communities, and you are at risk if the relationship fails.
>
> Don't let your community be comprised of too many colleagues who might disappear if you lose your job.
>
> Do have a mixed network of people of different ages and different employment situations and different family types. This way people in a group aren't competing for the same resources (like babysitters or a date).
>
> But don't be the only person of your type—like the only gay person or the only parent—because you will spend forever explaining your needs.
>
> Mix the intense and lighthearted. Have a solid central core of a combination of what Spencer and Pahl call helpmates, comforters, confidants, and soul mates, and surround that core with lots of associates, useful contacts, friends you turn to for favors, and fun friends.
>
> But too many friends actually becomes a drain, so have some friends who are both soul mates and fun friends. At the same time, relying on one person for all things is not stable, so make sure you have more than one of each type of friend.
>
> Don't rely too heavily on family—the healthiest personal communities are mostly friends, perhaps because friends are more accepting of your growing and changing—but if you need a lot of people around you with a sense of obligation and

responsibility toward you (like if you want help raising a child), be sure to include family or friends who are as committed as family in your mix.

Do have a mix of local friends and distant friends. Confidants can be distant but fun friends cannot.

Keep your old friends but add new ones, too. The healthiest communities evolve. They do not stagnate and they do not start afresh all the time.

Take an inventory of the types of relationships you already have. Do you have enough fun friends? Do you have enough confidants? Looking at the list above, ask yourself what is missing. Add those roles to the vision you have of your personal community and be on the lookout for people to fill them.

No Need to Reinvent the Wheel

You know what you want and you know what is missing. Maybe what you need has already been built and you can just plug yourself in. There is no need to go inventing new groups if there is one you might already like to be part of.

The question is, where are these groups? It used to be easier because the venues for interconnectedness were more visible—neighborhoods, clubs, and family groups. You could just join them. The new forms of interconnectedness are not quite so visible because they are based not on membership of organizations but on bonds of affection and shared values.

Go through your e-mail contact list and identify any friends you know who are really interconnected. Ask to join in on their community's activities so you can become part of it, too.

Tell your friends what you are looking for and ask if there is anyone they think you should meet.

Consider going to whatever events you have been invited to on Facebook.

If you want to be part of a save-the-world group, volunteer somewhere.

Use Meetup.com. As I write, groups meeting within two miles of my house include: pickup soccer, neighborhood LGBT twenties and thirties "non-sceners," ukulele players, and a real estate investment club.

Mining Your Existing Networks

John, a busy professional from Tennessee, wrote to me on Facebook and said he had realized that his problem was not that he didn't have potential connections. He didn't have to find new people. His problem was actually that he didn't really give time to and cultivate the ones he already had.

He told me he made a list of all the people in his life or just outside his life that had ever been important to him and whom he might like to be important to him. Now, some of us may not have a good list to come up with. But even if you come up with one, two, or three names, that's good.

So, make a list of everyone who has ever been important to you, who is important to you, and whom you would like to be important to you. You might go through your e-mail address book and your Facebook friend list to help prompt you with names.

Here's how to use the list:

To provide names of people you would like to build more interconnection with

To help you see what kind of people you have enjoyed being interconnected with so that you can build similar connections again

To help you feel nostalgic, reduce loneliness, and make you feel open to new connections (as we discussed on page 287)

Have Regular Gatherings

Research shows that people come to like and feel affection for those who are familiar. In one study, people were shown a random set of pictures of people's faces. The number of times each face was seen varied. The more people saw a face, the more they liked it.

For this reason, having a regular gathering lowers people's guards. It automatically builds interconnection. This is not to say everyone who comes to the gathering has to be part of your most intimate group. But it gives you a chance to meet people you might be interested in and for other people to do so, too.

Some hints for creating a regular gathering:

Create a weekly or biweekly event.

Aim to have between eight and fifteen people attend.

Choose a low-value night when people don't have a lot else to do, like a Tuesday.

If possible, choose a public venue instead of a private home, so people with weak connections to the group won't feel self-conscious about attending and so people can just stumble onto your group.

Choose a venue that is cool and nice so people want to go to it, but don't choose a place so popular that you can't push your tables together.

Invite more people than you want to come, say twenty the first time. You'll learn how many to invite as you go along.

Send an e-mail invite and then a reminder. Demonstrate high value. Mention the names of a couple of people who are coming and why they are fun. Maybe say that someone is going to perform.

Just because people don't come the first time doesn't mean they won't come the next.

Send an e-mail afterward to everyone—including those who didn't come—recapping all the cool things that happened. Remind everyone that the event will happen again the following Tuesday.

Add new people to your invitation list as you go along.

Focus and Select

So now you have a vision for your personal community, an idea of what kind of people you need and what roles you want them to fill, and a regular event where people are showing up and you can get to know them. Begin to focus in on people you especially enjoy doing things with—dinner, the movies, game night. Have smaller gatherings.

Or, to save time, ask them to join you in something you have to do anyway, like going to the Laundromat or grocery shopping.

Choose people who show up to your event night regularly. People who give back by helping to organize. People who have invited you places. People who want to be involved and help. These are the people who will commit to a personal community in the long term.

Maintain and Evolve

Ruchira Shah from San Francisco, who has a long-standing personal community, writes:

> The e-mail list is a big thing that helps us build and maintain our friendship network. Pretty much every day, I get

at least one e-mail from one member of the group. Sometimes I get just a couple in a day, other times I've gotten a hundred. Sometimes the e-mails are just recipes for dinner. Sometimes they are photos of children or pets. Sometimes they are silly anecdotes about the day. That's the majority of the e-mails. But, there are still a number of e-mails that go much deeper than that. E-mails about the big things: about fears, about struggles, about victories. The kind of e-mails that are too private and too complex to put on Facebook.

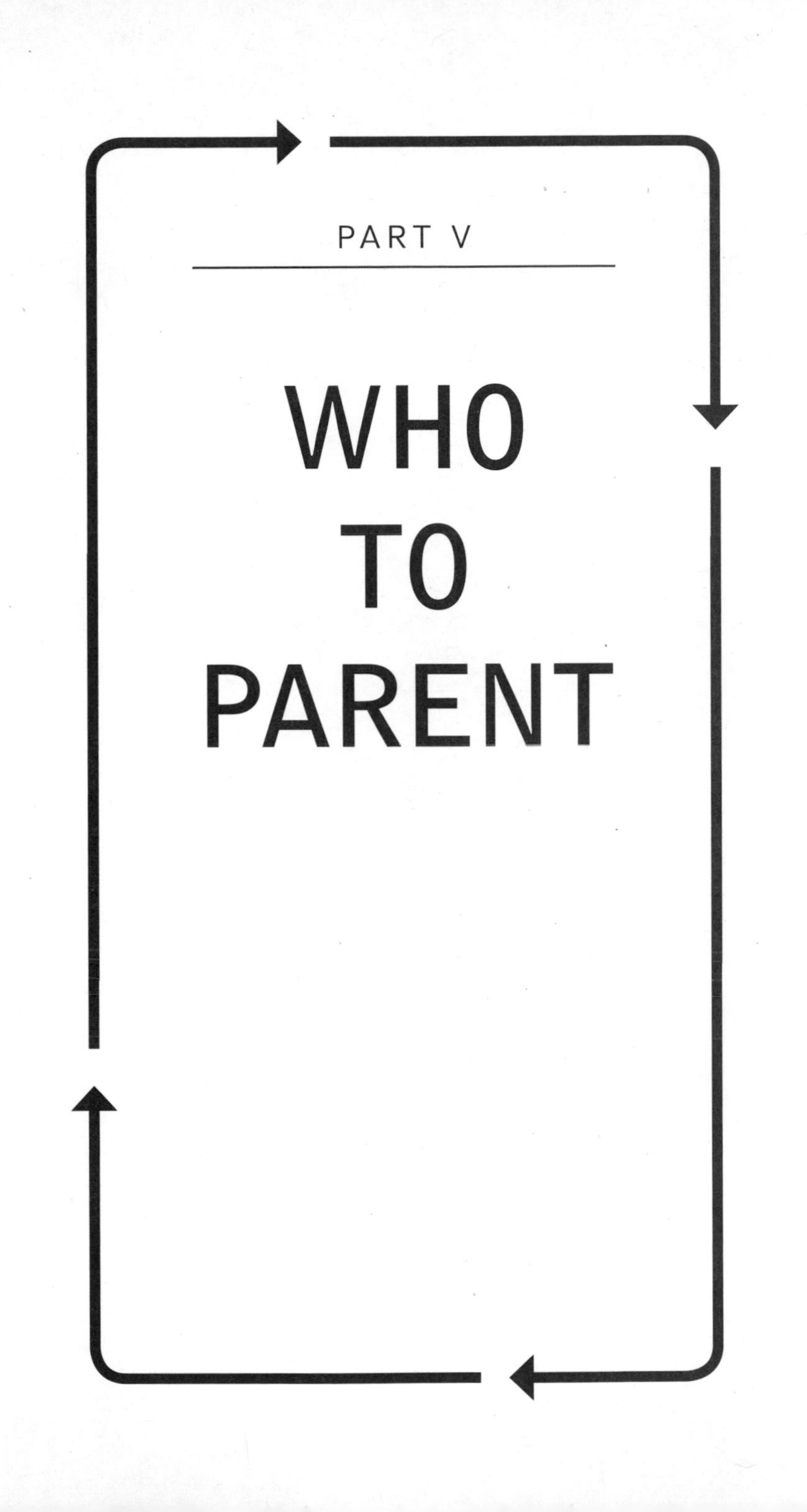

PART V

WHO TO PARENT

21

THE MANY WAYS OF PARENTING (INCLUDING NOT HAVING KIDS)

When You Pull on a Ball of String

He who does anything because it is the custom makes no choice.

—John Stuart Mill

Some lifequesters grow their own food or barter instead of buy or put relationship before career or bike instead of drive and on and on. Many of them begin by deeply examining one life choice, which tumbles them forward to the next life choice and then the next. Once you begin to really rebuild your life in a way that makes both you and the world happier, the standard life approaches start to unravel like a ball of string.

That is why it is no surprise that many of the lifequesters I've talked to about other life choices have also begun applying what they've learned to all sorts of decisions about parenting and having kids. Some lifequesters parent pretty much the way everyone else does—while putting a premium on being authentic and living in line with their values. Others find different ways to love children and other vulnerable groups without having kids of their own. Others have kids but invite other adults into their lives to assist with the children's upbringing.

Until recently, our society has defined three standard life approaches to parenting: having biological kids, adopting kids, or not having kids. Lifequesters, however, find all manner of other combinations. They broaden the definition of what it means to "parent"—to nurture and care for. They might, say, "parent" elderly people who need help. They "parent" youth in detention centers or adults in prisons. They "parent" children who want to learn music or immigrants who want to learn English. These are all valuable parenting choices.

We are told that the decision whether to have a child is our biggest decision and that there is no going back. But when you realize that the choice is really not whether to have a biological child but whether and how to embrace your inalienable role as parent, in the larger and wider sense, the pressure subsides. Because you can hedge. In a way, you actually *can* be a little bit pregnant.

Also, if the parenting of kids and youth is not defined by us each having our own biological children, then suddenly:

You can decide not to have kids and still a little bit have them by participating in the lives of other people's kids (we will discuss how later). You can decide to have kids and still a little bit not have them by sharing them with childless people (we will discuss how to do that later, too). In other words, you can have kids in ways that support people who don't have kids. You can also not have kids in a way that supports people who do. We would all be happier if child-free adults and adults with children helped each other.

For reasons we shall see:

> The biology of our brains actually demands that, for most of us to be at our happiest, we have to find a way to parent—even if it is unconventionally and not necessarily our own children or children at all.

> Who we decide to parent—and whether we choose to have our own biological children or not—may affect our life path more than any other decision.
>
> How we choose to parent may impact the world more than any other choice because it determines so much of what we leave behind.

Just as this isn't a self-help book, strictly speaking, it is not a how-to-parent book, either. However, it is a how-to-be-your-True-Self-in-a-way-that-helps-the-world book. Chances are, those of us who live authentically and refer to our Selves for guidance have at least some intuitive understanding of *how* to parent. That leaves only two questions: *Whom* should we parent? And how can we make that choice in a way that both supports people who have made different parenting choices than us and helps the world?

In this section we will talk about how to be our real Selves in relation to that question of who to parent. We will unpack the reasons people need to parent, the ways they make their choices about who they parent, what the world needs from our parenting, and the factors each of us needs to consider in order to make parenting decisions that are good for us and our worlds.

We Are All Already Parents

Disclaimer: When you start talking about the choice of whether or not to have your own biological children, the conversation gets loaded very quickly. So let's say this before we start: Choosing to have your own biological child can be a wonderful choice. Choosing not to have your own biological child can also be a wonderful choice.

Both choices are all the more wonderful if they are made authentically and pursued with true concern for Self and the world. That's what this conversation is about.

Anyway.

> *Instead of having clear prescriptions, it's better to work out the social structure so that people who never have children of their own will, in the normal course of events, help to take care of all the world's children.*
>
> —Erik Erikson, one of the founders of humanist psychology

It is worth remembering that, in Truth, we are all already parents to at least seven billion children—everyone on the planet. Some of our children are biologically related, others are not. Some of our kids are old and some of them are young. Some live far away; some live near. Some we haven't met and some we have. Some are born and some are not.

Whether we acknowledge it or not, we are all children to one another. This is a practical and spiritual truth. So, whether we have biological offspring or not, we can't escape being parents. The only question is whether we are willing to find our own authentic way to be *good* parents.

Don't let this overwhelm you. While we are all parents *to* one another, we are also all parents *with* one another. So we don't have to pull our hair out about all the kids. Some of us can worry about one or a group of the world's children and others can worry about a different group.

No matter how much we intend to give to the world in different ways, so many of us experience a very real pressure to have our own biological children. Our parents, our grandparents, our siblings, our friends all say a person should want to have children. A person should want to love and give to others. But why do they insist we have to love and give to people we have biologically produced?

Perhaps the standard life approach has become so individualistic that we think the only people we can really help and love and nurture are our own biological children.

Sometimes people call us selfish when we choose not to have children. But anecdotal evidence shows that people with children are much less likely to participate in the social change movement, for example, because of time constraints. So isn't it an equal good to choose not to have children so that you have the time resources to participate in helping others? What makes us selfish or not selfish is not whether we have our own children but the extent to which we use our lives to make others, as well as ourselves, happy.

Meanwhile, when you open up to the fact that all the world's children are each of our children, saying it is selfish not to have a child makes no sense, because you already have children. Maybe we could transform that cultural meme and put it like this:

Whether you have or will have a biologically related child or not, accepting your role as a parent to others is one of the greatest joys to Self and greatest services to the world.

This wider definition of the parenting role leaves us with an incredible amount of choice. You get to ask: *Which of "my" children call to me directly? Which ones do I want to parent most? How do I want to parent them? Do I want to parent children who come to me biologically or socially? Or both?*

We can parent by having our own babies. We can parent by helping, through our work and lifestyles, to make the world safer and happier. We can parent by taking care of children. We can parent by mentoring young adults. We can parent by helping older people.

There are as many ways to parent as there are people. What way of parenting is authentic to you?

As examples, other lifequesters have decided to:

> Raise their own biological children with an emphasis on citizenship:
>
> *I do have children. I have two sons. My husband and I have agreed that we do not want to have any more children. We love teaching our children about how to be responsible people in the world. We teach communication and emotional skills. We teach responsibility and*

ambition. Occasionally, I still have the urge to procreate. I know that we're done, but it still feels sad sometimes. I take care of that sadness by knowing we are doing what's right for the children we have. We talk about adopting or foster parenting in the future, and that helps me feel better, too. There are too many children who need a loving home that supports their future. The opportunity to make a difference for them is really inspiring to me.

—Meg Buck, Bethlehem, Pennsylvania

Raise biological children in chosen living arrangements:

I was a single biological parent of one, and I now share parenting responsibilities with a partner and my and her exes for five children. It helps us all live more balanced and full lives.

—Wendy Peterman, Albany, Oregon

Give service without emphasizing individual children:

I'm twenty-five years old, have absolutely no intention of reproducing, and don't have any hard feelings about it (maybe the hard feelings will come in the future?). I've been with my boyfriend for four years now, and we spoke about the topic of kids when we started dating. Thankfully, we were both on the same page, and remain that way. We really enjoy each other's company, we share many of the same interests, and we are both very physically active. We coach high school rowing as a girlfriend/boyfriend team during the spring and summer, and really enjoy working with the youth!

—Jenn McCallum, Ontario, Canada

Take care of other people's kids:

My husband and I have decided not to have children for a number of reasons. I work as a nanny, and have raised the same kids for six years. I love them dearly, but being a nanny has opened my eyes to just how much time, energy, money, and resources are required to have kids. Those are no small numbers. That, coupled with the knowledge of how many unwanted children need our help in the world, has impacted our decision not to have our own greatly.

—Catherine Chandler, Portland, Oregon

Give homes to foster children:

I'm in my late twenties and I have adored being around children since I still was one myself. I have a strong biological drive to have children, and I want desperately to be a mom. But there are many reasons why I don't want to have any biological children: impact on the environment, not wanting to pass down "bad" genes, fears that the pregnancy will go wrong, etc. But my most important reason is that there are already so many children who don't have good homes. Or any parents. My fiancé and I are already saving up money so that I can be a stay-at-home foster mom. And I hope that eventually we'll adopt some of the children we foster.

—Kasey Lindsay, Oakland, California

Do This!

Examine Other People's Modes of Parenting

Whether you are leaning toward having biological children or not, it will help you to break out of molds and find your own path if you start looking for alternative paths to parenting that others have found.

Begin to look around you and take note of other people's approaches to parenting and children. What are the many nonstandard ways you see people choosing to relate (or not relate) to children?

What are the ways you see people parenting biological children that could make your choice to have or not have a child happier for you and others? What are the ways you see people parenting nonbiological children that could make your choice to have or not have a child happier for you and others?

22

SOME REASONS PEOPLE DO AND DON'T HAVE CHILDREN

A Short Rationale for Considering the Biological Parenting Choice

As we have said, if the world is going to be a nice place to live, all of us need to parent—or nurture—more.

At the same time, there are so many people—children and adults—who need the time and energy we would give an as yet unborn child.

Also, having a child of our own is by far the biggest impact any of us can have on the planet, since each child can blossom into grandchildren, great-grandchildren, and so forth (having a child has 5.7 times more carbon impact than the total of everything else a woman does in her entire life).

That does not mean that everyone should have fewer kids.

Simply saying "no kids" or "fewer kids" denies fundamental truths of human existence, since some of us feel that we were literally born to have kids.

But *some of us* might actually be happier having no or fewer kids.

After all, according the U.S. Centers for Disease Control and

Prevention, three out of ten women report having given birth to a child they did not intend to conceive.

Which raises some interesting questions about the importance of making an active choice about whether we want biological children.

Philosophically speaking, isn't it true that, when a certain life course will impose a cost on our world community, we should not take that course lightly? That we should be sure it is what we really want?

To be clear, the question here is still not "*Should* I parent?" We've established that there is no escaping parenthood, at least as the term is defined in the wide sense of nurturing and caring for others. So the real question is "*Who* should I parent?"

Or to put it another way: Would it make me happier to nurture a(nother) child of my own, or would it make me happier to be unencumbered by child-rearing responsibilities and to nurture people—old or young—who are already born?

But since choosing whether to have a biological child is such a big decision, here is another thing to think about: How are we supposed to know for sure whether we really want a(nother) child of our own?

It may help to start by establishing some of the factors that affect people's choice whether or not to have a(nother) baby.

That's what this chapter is about.

Watching Out for Confirmation Bias

In discussions around the decision to have a baby, perhaps more than in any other conversation, people tend to give advice that reinforces their own decision—confirmation bias. *Confirmation bias* (also known as *myside bias*) is the tendency to search for, interpret, or remember information in a way that confirms one's own beliefs and choices. It is particularly strong for emotionally charged issues—like the baby decision.

People with kids feel guilty about saying they sometimes feel trapped, exhausted, or sad and so will only talk about the good stuff with nonparents. The bad stuff they will more openly share only with fellow parents—insiders who have made the same life choice.

On the other hand, those who have chosen not to have children in a society where having kids is still the norm have faced a lot of sanction. People can be so mean about other people's unconventional choices. So people without kids may not want to open up about any feelings of regret, either.

The baby decision is particularly fraught in our culture, and people can find it very difficult to differentiate their personal life choices from those of others. Be careful to understand people's biases when they discuss their baby choices.

Including mine.

Oftentimes, when people like me write about loaded social issues like the baby decision, they don't disclose their own personal choices. They marshal facts without offering the reader any chance to understand what personal bias the writer might have. That doesn't seem right.

So, fair disclosure:

Thus far in my life, I have chosen to have one child. As I write, Isabella is nine years old. She has dirty-blond hair and striking blue eyes. She calls me Dada and thinks it's embarrassing when I sing out loud on the street. I adore her and spend enormous amounts of time with her.

Yet I don't plan to have a second child, and it's not because of how mean-minded populationists say it's selfish to wreck the planet for your own gratification. It's because having one child is frankly enough for me.

Not in my darkest moments do I regret having my daughter. She is the love of my life. But I am still attracted to a less anchored life and I look forward to it (and maybe even inviting a grown-up Bella on some of my adventures). I look forward to a backpacking world tour. I look forward to being able to go on a long meditation retreat without worrying that my absence from the world will damage someone. I sometimes want to go to Africa to dig drinking water wells. But how could I do that if I needed to make money to support a second child?

For these reasons, as much as I am in love with being Bella's father, I will not reset my child-rearing clock to zero. End of fair disclosure.

Do This!

How to Learn from Others About Your Parenting Decisions

One of the best ways to figure out if a decision will make you happy is by talking to and looking at the lives of people who have made the same decision. Because of confirmation bias, you cannot just ask people if they are glad about the parenting decision they made. Psychologically, they need to reaffirm their choices and will probably answer accordingly.

There are two better approaches. One is to actually become a big enough part of people's lives to witness firsthand how their decisions affect them. The second, when you can't do that, is to ask questions in a way that won't be perceived as challenging to the rationale of the choice they have made:

What are you really enjoying about your decision?

What do you miss that you might have had if you made a different decision?

What is the best thing that resulted from your decision today/this week/this year/ever?

What is the worst thing that resulted from your decision today/this week/this year/ever?

What do you envy about people who made a different decision?

The Science of Why You Feel Sad When You Imagine Not Having Kids

Suppose you can see the benefits of not having a biological child to the life you want to lead. Suppose you are considering other ways to parent and other life plans that biological child-rearing wouldn't leave room for. Chances are, even if you have such thoughts, you still feel at least some tug to have a(nother) child. If you're veering away from that prospect, you may even have strong feelings of sadness and loss.

That's because our bodies, researchers have discovered, have two very real biologically prompted needs that drive us to have our own biological offspring. Most of us, whether we are male or female, experience these two needs to varying degrees.

The less common of these needs is experienced as an outright physical longing for a baby that has come to be called *baby lust* or *baby fever*. We will talk more about baby lust later. What we're going to discuss first is the more widely felt need—the biological drive to nurture creatures weaker than ourselves.

Our brains are hardwired by evolution to meet this need to nurture. A complicated system called the *nurturant bonding system* causes the release of chemicals like oxytocin and dopamine, which make us feel happy, when we behave in ways that move us toward fulfilling our need to nurture. That's part of why we love puppies and kittens and—yes—babies so much.

But if our brains perceive that we are moving away from fulfilling our need to nurture, ancient evolutionary circuitry drops the oxytocin and dopamine levels. More or less, your body starts screaming, "I need someone to take care of. I want the oxytocin rush that comes with helping little beings." This is why you may feel sadness and loss when you choose not to have a(nother) kid, even if you believe that all the facts point toward that being the best decision for you.

The Science of Why Even Parenting Other People's Kids Takes Away That Sadness

The good news—for both the individual and the world—is that the biological need to nurture can be redirected. Our first impulse, research shows, is to channel our need to nurture according to our upbringing and social circumstances. So, if everyone around you is having kids, or your parents have been nagging you for a grandchild, your first inclination when the need to nurture arises may be to have a(nother) kid.

But while the nurturant bonding system demands that we nurture, it does not specify the object of that nurturing. The system makes us desire to feel the warmth of affection for something weaker than us and the glow of satisfaction when we take care of it, but it does not require it to be our own biological child.

Thus, each of us can choose to meet our need to nurture in our own way. If you look around you, in your own community, you will find child-free people who manage to get nurture-triggered dopamine and oxytocin flowing in all sorts of ways—from volunteering for a structured child-care organization to formal mentoring to less formal relationships with a friend's or neighbor's kids.

One woman who volunteers to teach kids gardening and bike riding wrote to me, "Many parents think having their own bio kids is the only way to help future generations. They are not giving themselves enough credit. When I think back to my childhood, I think of the influence of the various mentors I had growing up, from teachers to Girl Scouts leaders and more recently college professors. These people had a huge impact on my life."

How the Rent-a-Kid Model Rocks for Everyone Involved

You might think at first glance that parenting other people's kids sounds like a consolation prize. Why not have kids of your own and get the oxytocin and dopamine highs by taking care of *them*?

Well, first of all, those nurture-based oxytocin highs cost bio parents between $300,000 and $500,000 per kid they raise, and that is only up until age seventeen. It doesn't include college. What we will call *social parents* get the buzz for free (or as close to free as they like, since many social parents spoil kids a little).

Second, it turns out that you don't get to fulfill your need to nurture as often as you'd think when you're bringing up your own kids. "Parents experience lower levels of emotional well-being, less frequent positive emotions and more frequent negative emotions than their childless peers," according to data gathered from thirteen thousand Americans by the National Survey of Families and Households.

According to author and Harvard psychologist Daniel Gilbert, careful studies of how women feel as they go about their daily activities show that they are less happy when taking care of their children than when eating, exercising, shopping, napping, or watching television. Partly, that's because parents are often too busy doing the laundry, getting stuck in traffic jams while driving kids from place to place, and working themselves into the ground to support their families in order to really engage in the nurturing behaviors that make parenting satisfying.

Also, full-time parenting can crowd out other things that make us happy—including, sometimes, other ways we can be of service to the world.

One mom wrote to me that she was inspired to have kids of her own because of the experiences she had working with other people's kids. But, she writes, "actual parenthood is not nearly as satisfying in certain ways. It is very hard for me to be that mentor to my own children, so preoccupied am I with simply the ins and outs of primary caregiving."

All of which leads to a paradox:

If you have a strong need to nurture, if you really want to have an influence on the lives of young people and see that you've helped them, you might find yourself more fulfilled using the time you would spend having your own kids—which requires nonmentoring activities like doing their laundry and working to support them—and dedicating that same time to the mentoring of other people's kids.

Meanwhile, since part of what reduces the joy of rearing children for a full-time bio parent is the busyness and the stress, having a nonbiological parent around to mentor and occupy your kid can lessen a little of the pressure on you and make parenting more enjoyable.

What if we all helped our friends and neighbors to parent their kids? What if those of us who have kids allowed neighbors and friends to help us? Isn't it just possible that this would be one way to support both people with children and people without children while making our communities and each other happier? Plus, we might all find we have the time to go to some parties and for the occasional run.

How Sharing Your Child Helps the World

When it comes to impacts on the planet of having a child, if you have already had one or some, you might think the discussion doesn't involve you. That it will be at best irrelevant and at worst guilt-provoking.

Not so.

People who have kids have a huge role to play in supporting those who choose not to. To wit, what if those of us who have kids invited help with nurturing from those who don't? Wouldn't it be easier for people to choose not to have children if they knew that they would still have a role in the lives of other people's kids? Wouldn't the extra support make it easier for those of us who do choose to have children?

Wouldn't that mean that people with children actually have a massive chance to help combat the planet's runaway population growth?

To understand the dangers of continuing population growth to our planet, think for a moment of a colony of brewer's yeast living in a barrelful of unfermented grape juice, soon to be transformed into your favorite chardonnay or sauvignon blanc. Brewer's yeast, if you didn't know, produces the alcohol in our wine, beer, and spirits.

Alcohol is the waste product left behind by yeast when it eats the sugar in grapes for wine, in hops for beer, or in potatoes for vodka. Alcohol is poison to yeast in the same way that human effluent is poison to us. Once there is too much alcohol in the vat of grapes, when yeast's little equivalent of a planet gets too poisoned, the yeast begins to die. Alcohol, in other words, is to yeast what pollution is to humans.

Now, think of what happens to a yeast colony when it encounters a yummy barrelful of crushed grapes. Suddenly, the population of adult yeast cells begins to grow and thrive. It feasts on the sugar, reproduces, and makes more yeast. After a brief *lag phase*, the population explodes. This is called the *exponential phase* of population growth, represented by the steep upward slope on the graph. It's a big baby-making party that seems, for a while, like it will go on forever.

But it can't. Because the barrel—like our planet—is a closed system. There is only so much food and so much room for the pollution. There comes a time when the adult yeast cells can't find enough food or start to get sick from the alcohol. Population growth slows down and then grinds to a standstill. This is the flat part of the population curve, called the *stationary phase*.

Ultimately, the yeast colony eats so much of the food and pollutes its environment so badly that individual yeast cells die off faster than new yeast cells are born. The population falls. This is represented by the downward slope on the graph. It is called the *death phase*. It pretty much spells the end of the colony.

Yeast, by the way, is one of the microorganisms scientists most often study to understand population growth and its consequences. A yeast colony follows what is called the *exponential population growth curve*. This curve demonstrates what happens to yeast in a closed container of crushed grapes, to a deer population eating its way through a limited forest, or to a population of human beings living on what, in galactic terms, is really just a small marble.

If you look at the graph of human population growth, you can see that humanity is, so far, on the exact same course as a typical yeast colony—it is following the exponential population growth curve.

Exponential Population Growth Curve

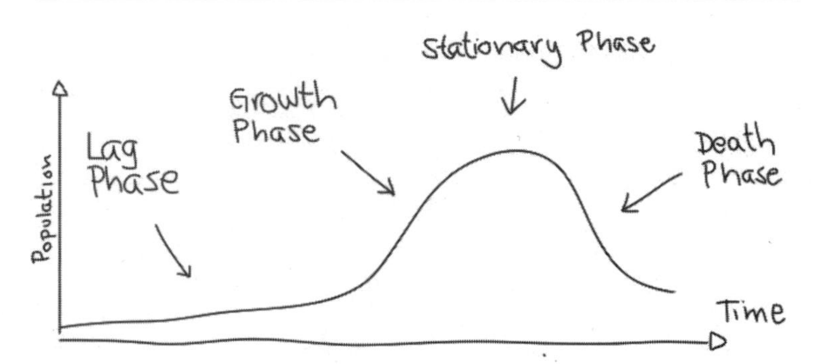

Human Population Growth

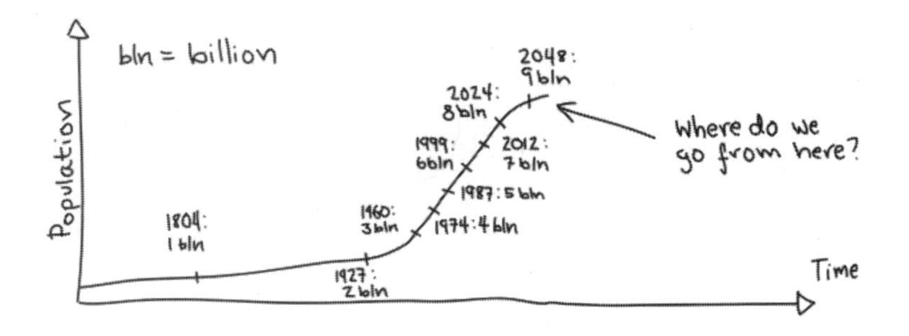

When humanity discovered oil and coal, we found, through the nineteenth and twentieth centuries, ways to leverage our work to produce a lot more food, medicine, and other cool stuff that extends our lives. Metaphorically speaking, we encountered our own version of a vat of crushed grapes.

Suddenly, lots of adult humans lived long enough to have lots of babies. Those babies, in turn, lived long enough to have lots more babies. It took about 200,000 years for our population to reach one billion in about 1804. But as we began to use oil and coal, we managed to increase our population by a second billion in only 123 years. The third billion came after 33 years, the fourth after 14, the fifth after 13, the sixth after 12, and the seventh—which we hit in 2012—after 13.

Like yeast in a vat full of crushed grapes, we have been reproducing like it's a party and having a bunch of kids who in turn use more stuff and create more pollution. For a while, it seemed like the party would go on forever. But it can't. Because the planet, like a vat of crushed grapes, is a limited container. The problem is the growing quantities of pollution in our land, water, air, and atmosphere. There is starting to be too much of it for us to live safely.

As a case in point, we could, of course, spend a lot of time talking about the various effects of climate pollution—the increased extreme weather events, the growth in deadly tropical diseases, the drought-caused famine. But it might be easier just to look at the fact that eighty thousand species are going extinct each year. As our planetary vat of crushed grapes is filling with pollution, its ability to support life is being strained.

The question is, how long before it starts having a hard time supporting human life? Looking at the growth curve graph, we see we may well be entering the stationary phase. How long before our population curve tips over the top and descends into that part of the curve that, biologists call the death phase? What will that be like? A plague? A famine?

One thing that could help us avoid that part of the graph is to slow our growth in population for a while. At least until we are able to implement renewable technologies that can meet our energy and other needs without demolishing the planet's ecosystems. We need to slow down and let the technology catch up with us.

Currently, United States population stands at about 320 million. The United States Census Bureau estimates that there will be about another 100 million of us—a total of 417 million Americans—by 2060. That population growth would cause a commensurate 30 percent increase in climate pollution—just when scientists are saying we need to *decrease* it by 90 percent, at least.

For these reasons, many ecologists would like the United States to flatline its population growth so that we would have about the same number of citizens in 2060 as we do now. Then, there would be no increase in climate pollution caused by population growth, which means technology and societal change would actually have half a chance of reducing the problems.

To achieve that doesn't mean all of us would need to stop at one or no children, but more of us would.

The good news is that, unlike those poor yeast creatures, human beings have the power of choice. Just because we have the resources to reproduce doesn't mean we have to. We have the ability to slow our reproduction before we completely degrade the planet's ability to support us. We can actually ask ourselves, *What would make me and my community happier? Is there a life path that includes fewer children but more life satisfaction?*

Dealing With an Ache That May Not Go Away

Many of us know people who want babies so badly that they literally cry when the subject comes up. They may talk about a tugging in their uterus or dream about babies when they sleep. Sufferers may weep with envy when they hear of other people getting pregnant.

This is not just histrionics. Because it turns out that, beyond the need to nurture, there is a second biological need that drives us to have babies: "baby lust," the unadulterated need to have a baby. Fortunately, it is far less prevalent than the need to nurture, occurring in only a third of women and a fifth of men, though it is also more demanding.

Researchers postulate that it may be caused by a vestigial surge of hormones that tries to drive people to procreate. Unfortunately, it can't be redirected like the need to nurture. In fact, it often returns almost immediately after a woman has given birth. In other words, even having a kid isn't a cure for baby fever.

Anna Rotkirch, Ph.D., director of the Population Research Institute at the Family Federation of Finland, who studies baby lust, compares it to romantic infatuation. She believes that there may be no way to make it go away, but that does not mean it has to be acted upon. "We don't necessarily follow our feelings in the case of infidelity, for example," she says.

One woman who chose to ignore the aching in her womb wrote to me: "I won't know what it feels like to be pregnant. I wanted to do that. I won't know exactly what it feels like to have that extra special bond with a little one. There is definitely a sense of loss. I get past that by keeping my goals in mind: I am on a path to being able to teach at a college level, and hope to someday be able to host workshops teaching troubled and homeless youth. I will give back to those less fortunate."

There are so many cases where our body's instinctual need for short-term pleasures—more food, more sex, more rest, more security, more babies—interferes with our longer-term ideals for ourselves. Detaching from these aches of desire is the subject of Part Seven, Rise Above the Noise. Because it's one thing if your hormones are pushing you in a direction you want to go, but it's another thing if they are pushing you where you don't.

23

PARENTING OTHER PEOPLE'S CHILDREN (AND LETTING OTHERS PARENT YOURS)

It's not about adults having kids. It's about kids having adults.

True fact: a child's having nonparental adults in her life can make or break her future.

In a classic study on youth, sociologists Terry Williams and William Kornblum write, "The probabilities that teenagers will end up on the corner or in a stable job are conditioned by a great many features of life in their communities. Of these, we believe the most significant is the presence or absence of adult mentors."

Experts suggest that some 19 million American children are at risk merely because they don't get enough adult attention. Whether we don't have our own biological kids or do, whether we plan to have more or don't, we can all do a lot of good if we give some attention, through formal or informal mentoring, to kids who are already born.

Youth paired with mentors are 46 percent less likely than their unmentored peers to begin using illegal drugs, 27 percent less likely to begin drinking, and 52 percent less likely to skip school.

Plus, remember how, back in Part Two: Wanting What You Really Want, we talked about how people who grow up with a lot of trauma and insecurity are less likely to become adults who value what is really important in life. They are less likely to become people who emphasize giving to the world because their childhoods imprison them in a cycle of fear and scarcity. Well, the more loving adults in a child's life, the more likely that process will be short-circuited.

Loving children makes a better world because when you love a child, the child is more likely to grow up to love the world.

Ways Adults Hang with Kids

There are many ways for child-free adults to spend meaningful time parenting others' kids: You can volunteer to teach swimming and hang with a group of kids with the added benefit of meeting other adults like you. You can volunteer to be a reading tutor in a school and create a more structured relationship with a smaller group. You can join a formal mentoring program like Big Brothers Big Sisters of America and get training and support from social workers. These structured models for parenting other people's kids are all excellent.

Finding a Tutoring or Mentoring Program

Formal mentoring and teaching programs are especially good if you:

Want to work with kids from a community you are not part of.

Haven't spent much time with kids before and feel formal support would give you more confidence.

Would rather work with groups of kids than single kids.

Are looking to offer specific educational or athletic skills.

Need structured boundaries between your work with kids and the rest of your life.

Good ways to find opportunities are to reach out and explain what you would like to do to staff at local public schools and departments of education, youth and children's centers, churches and places of worship, public libraries, YMCAs and other public sports centers, parks departments, and departments of social services.

Here are some national organizations that help connect adults with kids in need:

National Mentoring Partnership works to fuel the quality and quantity of mentoring relationships for America's young people in collaboration with more than five thousand mentoring programs in the United States. The partnership's mentoring opportunity tool lets you search for programs by area, age of children, type of work you are interested in doing, and type of relationship (individual or group). You can find more information at http://www.mentoring.org/get_involved/become_a_mentor/.

Reading Partners is a national education nonprofit that works in elementary schools to support students from low-income communities who are reading six months to two and a half years below grade level. They recruit and train community volunteers to work one-on-one with students for forty-five minutes twice a week, following a structured, research-based curriculum. You can find more information at http://readingpartners.org.

At-Risk Young Adults Engage in More Positive Activities When They Have Mentors

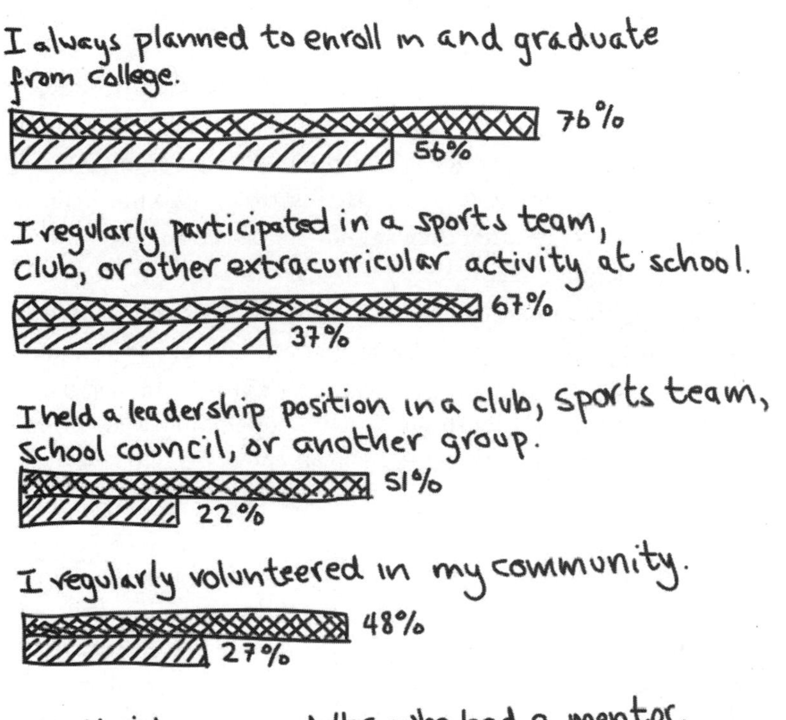

From a report by the National Mentoring Partnership.

But if you look around carefully, you will see that many adults—not just lifequesters—are *informally* involved in close relationships with children who are not related to them. Adults who love children sometimes just fall into long-term, committed, parenting-style relationships with the children of colleagues or neighbors or housemates.

A society where these kinds of relationships are the norm is a society whose children thrive.

Informally, these adults might be referred to as aunts or uncles or godparents. I like a term coined by a social work professor at Eastern Michigan University, Marti Bombyk, who studied these relationships: *social parents*. Social parents are adults who perform parenting roles to children they are not connected to by blood or profession but only socially.

The cool thing about social parenting is that it's an informal relationship that evolves naturally, and you can embark on it at any moment—with a colleague's kids, a neighbor's kids, or anyone else's. A social parenting relationship is based on opportunity, affection, and mutuality. It can grow naturally. And you can mold it the way you and the child want it depending on factors like how much time you have available.

We are going to emphasize social parenting because it points toward a way of being in the world: the idea that parenting can and should be part of all of our lives—whether or not we have children or are involved in formal mentoring programs. We are all already parents to one another.

The Benefits of Social Parenting

According to Marti Bombyk's study, the benefits of social parenting include:

To primary parents: free, responsible child care; support for parenting; parental satisfactions of seeing a child have a fulfilling relationship

To social parents: love and affection of the child, child's company, watching child grow and develop, playing with and teaching a child, rent-a-kid quality (meaning they can give the children back), participation in the world of parenting as a child-free adult

To children: having a confidant, additional adult attention, more emotional security, enhanced self-esteem, more recreational time, exposure to more adult lifestyles and role models, material benefits, having someone to spoil them

Social Parenting: Offering a Kid Something a Bio Parent Can't

Giving Understanding to a Transgender Teenager

Having adult mentors, friends, or Dutch aunts or uncles sometimes means kids have access to life experiences or ways of living that their bio parents just can't offer. Maybe the kid is particularly talented at music and needs a music mentor. Maybe the kid has diabetes and needs an adult friend who also has the disease.

Or maybe a kid is struggling with gender identity issues and needs an adult friend who has a special understanding. That was what Wendy Peterman of Albany, Oregon, was able to give when she befriended fifteen-year-old A.J.

The first time I met A.J. I came home and could hear that a bunch

of girls were up in my daughter Story's room. The door was closed and Story shouted, "Don't come in, Mom, we're getting changed." A few minutes later she called out, "Mom! Actually, can you come in? A.J. needs your help."

I went into Story's room, and there were a bunch of teenage girls in various stages of undress, and in the middle of them was this teenage boy, A.J. "A.J.'s zipper is stuck, Mom." A.J. looked horrified, like he was so busted. A boy in a dress with all these half-naked girls. He thought he was in so much trouble.

I didn't bat an eye. I walked up to A.J., straightened his dress, and did up his zipper. Suddenly every girl in the room was beaming because A.J. looked so happy. A.J., it turned out, was a drag queen who was just coming out. From that day forward, my house became a safe place for him, where he could be himself. I became an adult friend that he could talk with about his gender expression.

Let me take a step back.

Ever since my daughter was a small child, my house has always ended up being the place where all the kids come all the time. I just really liked having them around. So I'd take kids home when parents couldn't pick them up from school and always say yes when Story wanted a playdate. It got to the stage where other kids' parents began stocking my fridge for me because their kids were at my house so much.

So I have always had kids running in and out of my house. On the other hand, even though I kept the physical space for the kids, I never really imagined any of them wanted to be close to me. I was not very popular as a teenager and so I guess I kind of thought none of Story's teenage friends would really like me, either.

But with A.J. it was different, because of my life experience. I had had Story as a single mom but I married her stepfather when she was two and a half. When she was eight, my husband decided that he wanted to transition into being a woman. Though we ended up breaking up a couple of years later, we went through the transition together. The transition itself wasn't the reason we broke up.

So when A.J. stood there in my daughter's bedroom with a dress hanging off his back, my life experience meant it felt perfectly natural to me. After that, because Story was really attracted to A.J.'s ability to be intimate and vulnerable, he became one of her best friends. I was always driving the two of them around together and he always felt he could talk to me.

Also, it just so happened that he often took the same bus to school that I took to work and so we would sit together. We would talk about challenges he had with school accommodating his gender nonconformity. They were these short but intense conversations.

At the same time, his home life wasn't that easy for him. His mom had just left a seventeen-year Mormon marriage, was back at college, and was in the middle of coming out as a lesbian herself. She was doing a great job, but it was a time of crisis for her, and A.J. had to act pretty serious and responsible at home, helping to take care of his younger siblings.

At my house, though, he got to be a full-on precocious teenager. I provided a lot of stability and a place where he could be this hyper, over-the-top, hilarious, verbally inappropriate kid. Unlike at his mom's house, he had no responsibilities. He could just be a silly, immature, horny teenager.

What made A.J. so special to me among Story's friends was that he sought me out and had conversations with me, not just because I was Story's mom. He loved me as a mother figure. We had our own relationship. I felt I was able to be someone for him that no one else could be and that felt good. I even let him borrow and wear my clothes.

When I finally met A.J.'s mother, Stacy, we ended up going to see the play Rent *together, and as we were leaving I said, "Hey, could you please tell A.J. I need my black pants back?" She was like, "He borrows your clothes?!" It is one thing to have a friend who accepts your transgender son; it is a whole other thing to have one who actually lets him raid her wardrobe.*

Here's the funny thing: Stacy and I started dating. In July 2014, we got married. You know what made me so happy? When A.J.

made a toast, he said, "I'm not just happy for my mom. Wendy has adopted our whole family and really accepted us." I was lucky. I just happened to have the right life experience to accept and understand A.J. in a way that almost no one else could.

Do This!

What Kind of Relationship with Children Are You Looking For?

If you think you might like to have relationships with children who are not biologically yours, the first step may be to be clear about what kind of relationship you want. Take a pen and paper and let yourself imagine about what it could be like. Write it down. From your vision, answer the concrete questions below, which will help you determine your path forward.

What time commitment can I make?

What age of youth would I like to work with?

Would I like to work with one child or with a group of children?

Would I like to team up with other adults to mentor a child or a group of children?

What types of activities interest me? Do I want to help a youth learn a specific skill, pursue an interest, or succeed at schoolwork, or do I just want to be a caring adult friend?

What mentoring location would I prefer?

How to Become or Recruit a Social Parent

Explaining how to become a social parent or to recruit one is a little like explaining how to get a best friend. How do you do that? You can offer guidelines—like "Be open!"—but part of it is just about making yourself available. How do you do *that*? It's a little bit like fishing. You go to where you think the fish might be and drop your line.

One way to know where the fish might be is to think of where other people caught their fish. Here is a selection of ways that people from Marti Bombyk's Ph.D. dissertation began social parenting arrangements:

> A young couple moved back to their hometown to help a college friend who was becoming a single mom. They helped raise the baby.
>
> A fifty-five-year-old doctor befriended a mom across the street who was raising her child with her ex-husband. The doctor walked her dog every day with the little girl.
>
> A thirty-six-year-old carpenter chose to make a commitment to be an "uncle" to the five-year-old son of an ex-girlfriend and sees the boy several days every month.
>
> A thirty-two-year-old high school friend of a single mom on the East Coast stays once or twice a month for a few days with the eight-year-old daughter when the mom goes to see her boyfriend in California.
>
> A sixty-seven-year-old woman visited the hospital when her work colleague and his wife had their baby and has been visiting and helping as an adopted grandmother ever since.
>
> A twenty-five-year-old woman rented a bedroom from a couple who had a young daughter. She no longer lives with them but maintains a regular, sisterly relationship with the twelve-year-old.

A thirty-six-year-old psychologist decided she does not want children. She has Sunday lunch with an old high school friend, his wife, and their two children. After lunch, she takes the two kids for walks in the woods or out to buy stickers they collect.

A twenty-six-year-old master's student accepted a college professor's invitation for Christmas dinner when she couldn't go home. After that, she started babysitting for his children and loved them so much that she stopped accepting payment.

A thirty-one-year-old accountant's dog attracted the interest of three siblings who lived down the block. The children began to visit regularly and now the accountant is friends with the mom and is like an uncle to the children.

Social parenting arrangements are as varied as people. Some include nearly daily contact. Some are regular but not daily. Some are occasional. Sometimes the contact is entirely recreational and sometimes there is child care involved. Many times the relationships last for a long time—years. But the arrangements are diverse and suit everyone involved.

None of the examples point to a specific way of being a social parent. But they point to the ease with which they were formed. Who among us doesn't have neighborhood kids who like dogs or a favorite college professor who might invite us over for Christmas? The circumstances may be unique but the general tenor that is common to them all is openness.

In her dissertation, Marti Bombyk offers suggestions for people who want to be social parents:

Find the kids in your network. Do you have friends with kids? Or colleagues? Or are there kids in the neighborhood who seem interesting? Look for kids whose parents would be open to the social parenting relationship and maybe have a particular need for the kind of support you can offer.

Develop trust with parents. Don't initially seek to spend time with a child alone, but make a date with the parent and child together. Establish a relationship with the family. Once your trustworthiness is established, then you can offer to babysit or to take the child for a short outing.

Move slowly. Your offers should start small. Don't offer to take the child for a weekend, say. Everybody—you, the child, the parents—needs to get used to the new relationship. Let everyone get comfortable and confident with small occasions first.

Follow the primary parent's rules. Learn to understand the rules the parent wants you to follow with the child. How much TV or video gaming is the child allowed? What foods can they eat?

Make the primary parent's life easier. Figure out how to make it so your relationship with the child enhances rather than disrupts the primary parent's life. Your social parenting will last longest if it supports rather than drains the family.

Bombyk also offers guidelines to bio parents who feel they and their children might benefit from a social parent:

Identify the right type of person within your network. Successful social parents live nearby, are often playful and young at heart, are generous spirited, and are very clearly child oriented.

Socialize with your child. Give potential social parents the chance to meet and enjoy interacting with your child. Invite them for dinners or other outings. Let a relationship between them establish itself in your presence.

Let your trust be known. Let the potential social parent know that you would allow him to pursue an independent relationship with your child. Maybe they could naturally go to a movie or a museum they share an interest in but you do not.

Enhance the social parent's life. Try to arrange the relationship so that it is convenient and does not burden the social parent's life. Get clear on how much time the social parent would like to dedicate and what your mutual expectations are.

Appreciate the social parent. Make it clear that her role is important to your child and valued by you. Treat the relationship with respect and honor arrangements. Make her feel part of the family and that the lack of a blood bond is not important.

Social Parenting: Sometimes the Best Way to Be a Social Parent Is to Adopt a Whole Family

So many people wish they could have children who can't—because of either biology or other life choices. So many parents need help. So many children don't get enough adult attention. Instead of adopting babies, why don't we adopt families? Share the parenting. Share the children. Share the love. Judy Green, from Ruther Glen, Virginia, told me her story:

We live in a planned community with sidewalks and a pool and a clubhouse. We were at the pool and this woman was doing laps and she had left her little girl napping in her stroller. I admired the mom for taking care of herself, but I couldn't help keeping a protective eye on the little girl. Some people—like me—are like that. The parenting urge is so strong.

Anyway, when the mom—Sharon—got out of the pool, we chatted. Not too long later, we were all at bingo night in our community: Sharon and the little girl, Lesley, and me and Joey, my twenty-nine-year-old son who has Down syndrome and lives with us. We all struck up a friendship and we started visiting each other's houses.

I kept offering to watch Lesley for Sharon when Sharon was having health problems. She was too polite to say yes. But one day I went to visit and little Lesley's hair wasn't brushed. I knew Sharon must really have felt poorly. I told her to rest and I would have Lesley come visit. That was the beginning.

Since then, Lesley comes and spends time with us during the day all the time. We just stay in touch by text and sometimes I just say that I was thinking of coming to get Lesley. Sometimes, Joey goes over there, too. Sharon is his friend now, too. For me, it's like having a granddaughter. We are like family.

24

DECIDING WHETHER TO BE A BIO PARENT

The only right is what is after my constitution; the only wrong what is against it.

—Ralph Waldo Emerson

When issues are complex on the outside—you are faced with a mess of social *should*s and *shouldn't*s—often the only way to navigate them is by looking for guidance from the inside. By accepting that the facts don't seem to be pointing to a clear direction and looking instead to your own wisdom. By asking, *What is my own path? What is my life for? What is my True Direction?*

If you are clear that a central purpose of your life is to have a(nother) child, then you probably should. Similarly, if you are clear that you don't want a child, then you probably shouldn't. But for some of us, the decision is not that clear.

Merle Bombardieri, author of *The Baby Decision: How to Make the Most Important Choice of Your Life* and a social worker and coach who has been helping people make the baby choice for twenty years, points out that in our anxiety about the decision to have or not to have a biological child of our own, we sometimes fail to make a decision at all. We are scared of what we want and so don't acknowledge it.

Bombardieri writes that Abraham Maslow, the late psychologist best known for his hierarchy of needs, distinguished two types of motivation: growth and deficiency. Deficiency motivation is a mechanism by which we find ways to avoid what we worry might be unpleasant. Growth motivation is a mechanism by which we figure out how we become the people we are meant to be.

Bombardieri points out that people may decide both to have their own children and not to have children for reasons of deficiency. Some people halfheartedly avoid having children and eventually have an "accident" because they can't bring themselves to face the disapproval or fear of making a nonstandard decision. Having a baby isn't a life-affirming or growth decision for them but a way to avoid the discomfort of making a decision that seems frightening. Later, they might feel like the victim of their "mistake" and never have to admit they betrayed themselves.

Similarly, some people don't actively decide not to have a baby but instead postpone it and tell people that the time isn't right yet until the moment passes. Again, they have avoided the discomfort of making a decision to be child free or to face the commitment that having a child takes. Their nondecision to not have a baby is not life affirming. They throw themselves into careers as a kind of cover.

Making safety decisions means that we have to find ways to get away from the truth of our decisions. Because we don't acknowledge them, we miss the possibility of asking how we want to follow the path in a way that fills our entire need. How can we not have children in a way that acknowledges that we still want to nurture? How we can have children in a way that acknowledges that we still want some freedom?

On the other hand, people acting from growth motivation examine their life choices and consciously come to terms with what they want. They acknowledge who they are and take the risk of following that path. By knowing and coming to terms with who they really are, in the case of the baby choice, they are able to have kids or to not have kids in a way that makes them happiest and most helps the world.

Being positive about who you are and finding your own path forward is the way of the lifequester. Lifequesters both decide to have biological children and decide not to. What makes the choice authentic is not which they choose but *that* they choose.

What makes you a lifequester is not whether you have your own kids or adopt them or don't have kids but whether you make your choices in ways that are authentic to you and help the world. What makes you a lifequester is looking at your needs and the needs of the world around you and asking what the path forward is for you.

You don't have to be "alternative." All that makes you a lifequester is that you actively choose what is authentic to you.

Are you alive to who you really are? Are you awake to the world around you and its needs? Do you do things because they are what everyone else does? Or do you do things because you are awake and conscious and want to do the best by yourself and everyone else?

For all our talk, if you are struggling with the decision of whether to have a(nother) kid, there is one crucial piece of information that you still may not have: how you will feel once you've made the decision. This piece, according to Merle Bombardieri, is crucial.

If you are not sure, social parenting is one way to begin to discover how the responsibilities of child-rearing fit your life. It is also a way to see whether it is close enough to biological parenting to be enough for you.

Of course, social parenting isn't the same as being a bio parent, and it doesn't let you know what bio-parenting would feel like. To help with making the decision, Bombardieri suggests projecting ourselves into various alternative futures through visualization and seeing how we feel. There are many exercises to help with the baby-making decision. We will discuss two adapted exercises that Bombardieri recommends in her book.

In both cases, grab a pen and some paper and write down your reactions. When you do the exercises, remember that you can make a meaningful contribution to posterity whether you have a(nother) child or not. If you have a strong nurturing instinct, remember that you can mentor other people's kids even if you have none of your own. If you have a strong desire to do service to your community, remember that having children doesn't mean caring for them will be the only thing you do with your life.

The Gravestone Test

This exercise will help you decide whether a life with a(nother) child or without one is more in line with the future you want.

First, assume that you have not had a(nother) bio child. Close your eyes. Imagine the events following your death. What will be inscribed on your gravestone? What will people say about you at the wake? Who will be the most important mourners? What will you be remembered for? Let yourself really fantasize different scenes. Listen in on conversations. Open your eyes and take some notes on what you've imagined and what feels good and bad.

Now, assume that you *did* have a(nother) child. Close your eyes again. Do exactly the same exercise. Imagine the funeral. The obituary. The memories people share about you. Open your eyes and take some notes.

Which did you like better? Why? What were the elements of each that were good? What were the elements that were bad? How does this affect your decision whether to have a(nother) child?

Your Future Parenting Photo Album

Being a parent contains a whole range of satisfactions and dissatisfactions.

For each of the scenarios below, close your eyes and imagine it happening to you (or, where relevant, your partner if you are not the birth mother). How do you feel? What do people say to you?

What do you say to them? How is your life different as a result?

Write some notes on how you feel after you've imagined each scenario. Were you able to embrace all the images? If you become a parent, are you willing to accept those bad moments you may have wanted to skip over? If you don't, are you willing to relinquish the pleasures?

> You or your partner is pregnant. There are body changes. Perhaps there is nausea some mornings. The tummy grows. Imagine beginning, middle, and late pregnancy as your body changes and your capabilities differ.
>
> You have your first ultrasound and discover the baby is exactly the gender you wanted.
>
> You have your first ultrasound and discover the baby is not the gender you wanted.
>
> You or your partner is giving birth.
>
> You bring home a wailing, red-faced newborn.
>
> You relax in the living room while nursing a peaceful three-month-old.
>
> Your baby cries and cries in the middle of the night and nothing seems to comfort him.
>
> Your one-year-old picks up her spaghetti, pours it over her head, and smiles.
>
> Your toddler takes his first step, falls, and looks surprised.
>
> Your toddler has a tantrum in the supermarket, kicking and screaming.
>
> Your three-year-old dances in the moonlight.
>
> Your five-year-old "reads" a book to you, making a story up to go with the pictures.
>
> Your seven-year-old tells you that if you loved her, you would stay home like other moms/dads.

The school you found turns out to be making your child terribly unhappy.

Your nine-year-old brings you breakfast in bed.

Your eleven-year-old wins a prize in the school science fair.

Your fourteen-year-old refuses to do as you say and shouts, "You can't make me!" loud enough for the neighbors to hear.

You come home after leaving your sixteen-year-old alone for the night and the house stinks of pot smoke.

You look out your bedroom window to see your child making out with someone on the street.

Your seventeen-year-old asks your permission to go on the pill.

You get your child's first-semester bill from college.

Your nineteen-year-old decides to take a year off.

Your twenty-two-year-old wants to live back at home for a while.

You look at the life of a young adult you admire and love and realize you had a crucial role in creating it.

We've said a lot, so here's a road map to considering the question of whether having a(nother) child suits you and the community you live in, at least according to the arguments discussed here:

Caring for kids is a huge pleasure for many people. It is central to many of our lives. Central to our very purpose. Yet the planet can't stand for us all to have children at the rate we have been. At the same time, there are some 15 million children in the United States alone who need more adult attention.

This does not necessarily mean we should sacrifice our own happiness. It does mean we should be clear. We should not stumble toward a fate that makes neither us nor our planet happy.

If you want kids, have them. If you don't, don't. Those who aren't sure, examine yourself. You might begin by getting clear about your True Direction.

Ask yourself: Why do I want kids? Is it at all because I feel expected to by someone else? Is it the need to nurture? Can I nurture in another way? In a way that is more satisfying to me?

Is it baby lust, a chemical reaction happening in my brain? Do I want to be ruled by it? Is that chemical trying to force me to do something that might have been right for an ancient tribe but is not right for me? Or is it forcing me to do what I actually want to do?

Do the visualization exercises above.

Stop telling other people that they should have kids. Let's make not having kids an acceptable life path.

Help people who already have kids.

If you have kids, accept help. Share them.

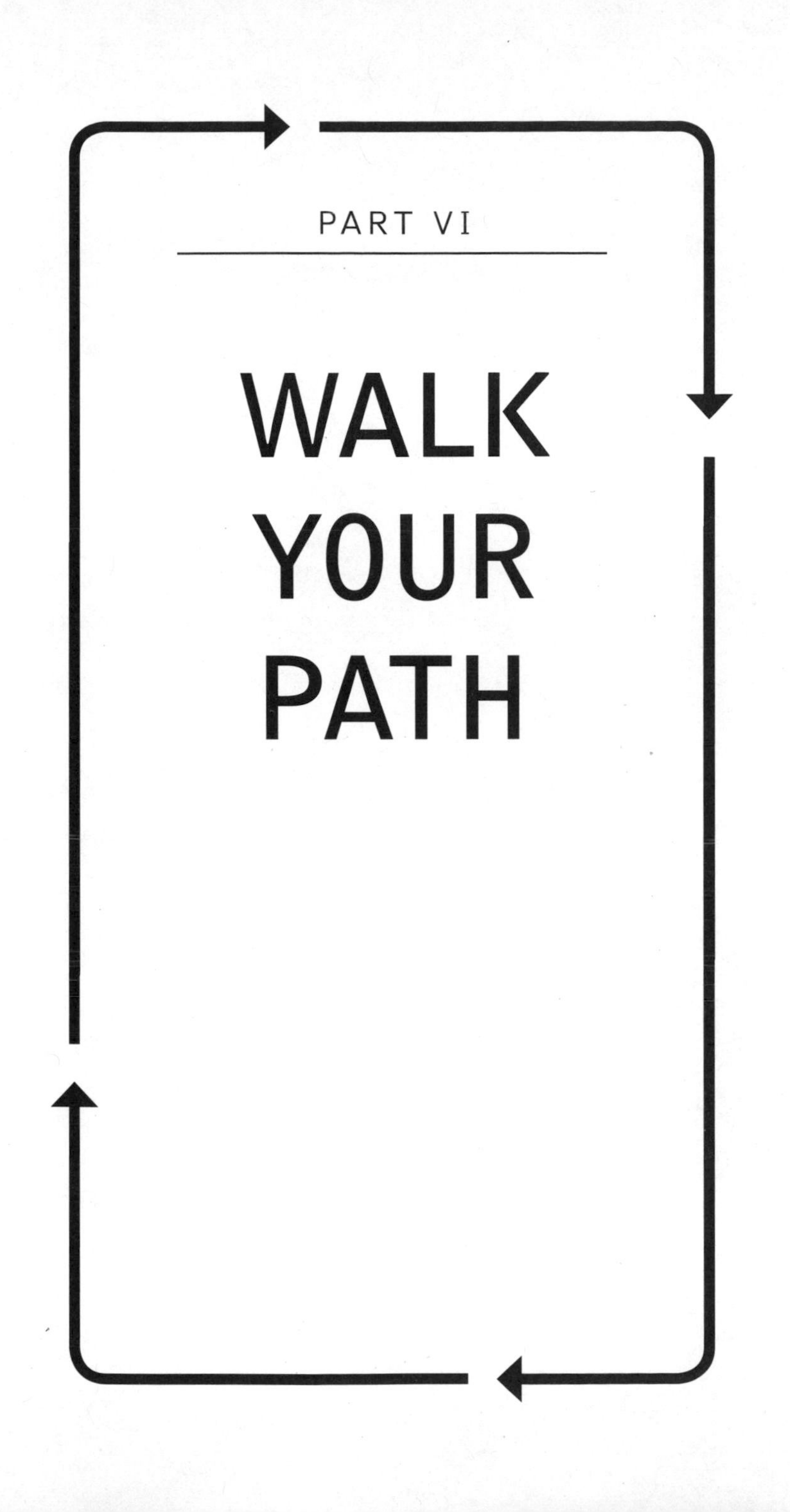

PART VI

WALK YOUR PATH

25

LEARN TO DISCERN

Experiments in Truth

Douglas Steere, a Quaker teacher, says that the ancient question "Who am I?" inevitably leads to a deeper one, "Whose am I?"—because there is no identity outside of relationship. You can't be a person by yourself. To ask "Whose am I?" is to extend the question far beyond the little self-absorbed self, and wonder, who needs you? Who loves you? To whom are you accountable? To whom do you answer? Whose life is altered by your choices? With whose life, whose lives, is your own all bound up, inextricably, in obvious or invisible ways?

—Reverend Victoria Safford,
author of *Walking Toward Morning: Meditations*

We have come a long way since the beginning of this book. We have talked about how we are not powerless to change our lives or the world. How we don't have to be helpless like Ralph the dog (or Ralph the accountant). We have talked about the failure of standard life approaches to create lives of meaning and purpose and how Truth lies within each of us.

We have talked about how the path of the satyagrahi—the Truth follower—leads to happiness in our own lives and in the world. We have talked about little ways to follow Truth and build meaning and purpose through our smaller life choices. We have learned to build community and to parent. These are our experiments with Truth.

With any luck, if you have taken any of the actions we have discussed, you have begun to experience a shift within yourself. You may actually, at times, have begun to experience some physical sense of centeredness, perhaps as a feeling of strength in your chest or an anchoredness in your feet. Hopefully, you have developed some faith in the process which is You.

If you have, you may also have begun to find that you want to go further, like a fruit ripened on the tree, finally ready to fall off. You may be at a place where you can really ask what your vision is. Who are you? What is your calling? We have discussed changing things at the periphery of our lives. Some of us may now want to reconsider the focus of our lives.

Perhaps the most common standard life approach is to consider the focus of one's life to be career. We are not going to talk about career. We aren't going to talk about what we want to do but who we actually are—not just in our work but in all areas of our lives. We are going to talk about our calling, our vocation. Not what we were put on this earth to do but how we were put on this earth to be.

The discussion of calling and vocation is a long one and we can't cover it all here. But we will begin. We will talk some about how to *discern* the way you are called to be.

We will talk about bringing our use of time—the real currency of our lives—in line with our True Self. This will bring us to a new way of thinking about money and finally another way to think about work. We will look at money and career as important only inasmuch as they help us get value from our real commodity—our time.

Thus, we continue our experiment with Truth.

Go Where You Are Called

We get scared that being ourselves won't get us where we want to be, but if being ourselves gets us there, then that is *where we want to be.*

When people attached to standard life approaches tell would-be lifequesters to "be realistic," they often use life's limitations to bolster their argument. *Everybody has to pay rent. Everyone has commitments. You can't just be yourself. You have to keep your promises, fulfill your responsibilities. You can't just follow whatever path you want to follow!*

Of course, this is true. You can't just follow whatever path you want to follow. But the lifequester knows that this is true for a different reason than most people think.

A lifequester follows Truth. That means finding a path that is authentic to the lifequester while responding to conditions in the world. A lifequester tries not to betray either herself or the world by, say, choosing a career path that is not True. A lifequester knows better than to go into the supposedly more lucrative sciences to be "realistic," say, if she has the temperament of a poet. Or vice versa.

Because even if you do not embrace the lifequest, if you choose a path that is not True, the forces of those parts of your Self that you have tried to bury in your shadow may eventually force you off the false path. Then, think of how many broken promises there will be. Think what happens when you finally leave a career or marriage that you didn't really want and you are forced to decommission a life you have already built.

This is why the lifequester chooses her commitments and responsibilities carefully. She does not choose the ones that are in line with "getting her what she wants." She chooses the ones that are in line with "being who she is." In this way, she keeps her promises and honors her commitments not because she said she would, but because they are in line with her nature.

This is what we mean by "going where you are called."

Then, Build Your House Where You Live

Imagine a farmer who has captured a strong wild horse. Other farmers have tried to corral this horse, but every day the horse just jumps their fences and every day they have to fetch him back from the surrounding countryside. Each of them has given up, but now our new farmer has a turn.

Each morning, he feeds and waters the horse in the stable and then opens the stable door so the horse can roam free. Each evening, his fellow farmers come to him laughing. "Your horse is in my field" or "Your horse is on the hill." Our farmer goes and gets the horse, brings him home, beds him down for the night, and begins again in the morning.

Over time, the horse roams the valleys and hills and farms. But one night, a friend tells our farmer, "Your horse is under the tree by the stream." And a few nights later, the farmer again finds the horse by the stream. Before long, no one needs to tell the farmer where the horse is. He is always under the tree by the stream.

The wild horse has explored the land far and wide and has found the place it likes to graze and rest. Now our farmer gathers together the fence and posts for his corral. Everyone says it is pointless. The horse will jump the fences, just as he always has. Except the farmer erects his corral under the tree by the stream. The horse never even attempts to escape. The other farmers are amazed.

One day, our farmer knows, his horse will have eaten the grass down or the wind will blow differently and the horse will jump the fence and begin to roam. But the farmer does not worry. The horse will let him know where next to build the corral.

Let yourself wander. Then build your limits where you find yourself staying. That is the place from which you will most comfortably and happily serve the world. That is the way to build your house where you Truly live.

26

HOW CAN I HELP?

Outside Job, Inside Job

One day, the late Zen Master Seung Sahn, the Korean founding teacher of the Kwan Um School of Zen, in which I happen to be a dharma teacher, sat down at the Providence Zen Center in Rhode Island to give a talk.

He said, "Money, sex, fame, food, and sleep are part of desire mind." By saying these things are part of desire mind, he did not mean they don't matter. We are born in a human body and so we need human things. Money, sex, fame, food, and sleep—in the right measure—can facilitate the Good Life. "These are important," the Zen Master went on.

"But what is *most* important?" the Zen Master asked. "When you get old, then you will understand. I am sixty-six. When you get to be my age, desire disappears: sex mind disappears, fame mind disappears, and sleep mind also disappears. All desires disappear. When that happens to you, what will you do?"

We know that money and sex and fame and food come and go. That in itself is enough to make many of us question whether we should really orient our lives around chasing those things. But on top of that, the Zen Master said, even the desire for those things disappears. Without those desires, what is the direction of your life?

"Many people say that a human being's job is to get money, sex, et cetera. But that is only a human being's body's job—an outside job. What is your inside job?"

That is the big question of our lives. Since not only the things we desire but the desires themselves disappear, what is your True Direction?

What is your inside job?

The Choice of No Choice

What Zen Master Seung Sahn calls our inside job, a Christian might call "vocation." Industrialized society has appropriated the word *vocation* to mean a career or job you are trained to do. But the word comes from the Latin *vocare,* meaning "to call." In the True sense, vocation means that which you are called to do.

When we say "that which you are called to do," however, we don't just mean what you are called to do in work but in all situations. You don't need to have a job to have a vocation. Vocation is a life path, not a career choice (though it can, of course, affect your career choice). So a better definition of vocation than *what you are called to do* is *the way you are called to be*. In those quiet moments, when you are without fear or overwhelming desire, what moves you to act?

When people hear of vocation, they sometimes worry. They conflate the idea with the expression "to do God's will." To many people, to do God's will means to suppress one's own will. To try to do my inside job to the detriment of my outside job. To sacrifice ourselves for the sake of some other.

I once heard the spiritual teacher and author Marianne Williamson say, "Some people worry that if they do God's will or follow His calling, then they will be asked to be an accountant when they really want to be a musician. But what kind of God would ask a person with a passion for music to be an accountant?"

Doing what we are called to do does not mean ignoring what we need or even ignoring what we want. In the story of Zen Master Seung Sahn above, he said that our outside job—fulfilling our needs and desires—is important. Then he asked, "But what is *most* important?" Or to put it another way, what is the reason for bothering to do our outside job? Why bother to fulfill our desires?

We do not ignore our outside job, but we recognize that we do it partly because it helps us do our inside job. The object of getting money is not to get more and more money until you are rich and lonely. The object of eating is not to eat more and more food until you are sick and fat. In part, we do our outside job in order to keep us happy and healthy enough to do our inside job.

When we are aware of our inside job and we understand how we are called to be, we fulfill our desires:

1. In ways that help other people.
2. In ways that give us the energy to do our inside jobs.
3. In ways that give us enjoyment.

Just as we have an inside job and an outside job, so there are also two elements of vocation. Vocation—the way you are called to be—is partly an expression of the prized skills and talents and passions inside you, who you are, because only that which is inside you can be called from you. Vocation is also your answer to what the world needs or calls for that most resonates with you.

Remember the needs to feel autonomous, effective and related that we discussed in chapter 8? True vocation fulfills those needs. Thus, none of us is called to be someone other than who we are to help the world. We are called to be our Selves to help the world.

What is called from you—how you might be—in response to a catastrophic earthquake, say, depends on who you are and what aspect of the tragedy most moves you. If you are a musician who cares for children, you might be called to visit the hospital children's ward and play songs with your guitar. If you know first aid and you have an affinity for elderly people, then you might climb the stairs in tower blocks that have lost power and tend to their injuries.

As the Christian writer Frederick Buechner put it in his book *Wishful Thinking: A Seeker's ABC*, vocation is the combination of (a) what you most need to do, and (b) what the world most needs to have done.

He wrote that if you really dig your work as a writer, say, you have probably met requirement *a*, but if you are writing TV commercials for deodorant, you may have missed *b*. On the other hand, if you are a doctor working in a leper colony, you have probably met requirement *b*. But if it depresses you and makes you unhappy, you have missed requirement *a*.

"The place God calls you to," Buechner writes, "is the place where your deep gladness and the world's deep hunger meet."

Do This!
When You Went Where You Were Called—and When You Didn't

To begin to understand how you are called to be, spend some time writing about those periods in your life when you felt you lived in line with your "deep gladness" and the world's "deep hunger"—and when you didn't.

For four consecutive days, spend fifteen minutes writing about how you were spending your time and how you felt about it when you were living in each of the quadrants in the diagram that follows.

The Four Ways of Being

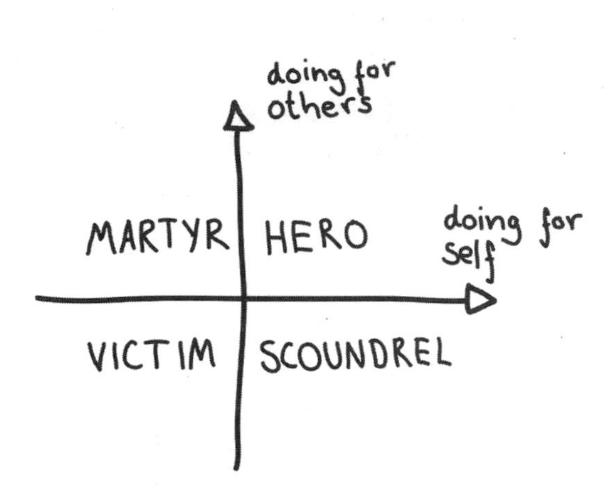

Hero: When were you both having the time of your life and feeling like you were really helping the world? When did you feel kind of heroic because all you were really doing was what made you happy but it was good for others, too? What were you doing? How did you feel about this period compared to other periods? What aspects of this period should you definitely strive toward in being the way you are called to be?

Scoundrel: Write about a time when you felt you were riding high on the hog but perhaps you weren't really helping anyone else. How did this period feel compared to living in the other quadrants? Is it a place you would be willing to stay long term? What were the really nice parts about it that you would like to keep if you were able to help other people at the same time?

Martyr: Write about a time when you may not have been particularly happy but you did feel you were being helpful to others. What kind of help were you giving? What satisfied you about it? What satisfactions would you like to maintain if you also managed to glean a little more pleasure for yourself?

Victim: Write about a period of your life when you felt you were fulfilling obligations that brought you precious little happiness and didn't help anyone else, either. What brought that feeling about? What were the qualities of that time? What concrete aspects of that period should you avoid in the future?

When you are following your calling, you will most likely live in the hero quadrant—helping yourself and others. But what does information from living in the other quadrants tell you about how you, in particular, should live in the hero quadrant? What talents would you be using? What would you be doing for fun? Who would you be helping? How would you be helping them?

Letting Your Self Decide

Therapist Richard Schwartz, Ph.D., developed the Internal Family Systems model of psychotherapy and founded the Center for Self Leadership to promote letting our Selves lead us through our lives. Back in the 1980s, he began noticing that several of his clients described extensive internal conversations between different parts of themselves when he asked the clients to discuss their problems.

One of these clients, whom Dr. Schwartz calls Diane in his writings, described a voice that always told her that her life was hopeless. Schwartz had Diane ask the voice why it said she was hopeless and it replied that it said Diane was hopeless so she wouldn't take risks and get hurt. The hopeless voice was trying to protect her.

If that pessimistic part of Diane really wanted to help her, Dr. Schwartz reasoned, then maybe Diane could negotiate a different approach with the pessimist inside. But Diane refused. Instead, she went on a rampage about how she just wanted to get rid of the pessimistic part because it made every decision in her life a major hurdle.

Diane had previously spoken of battles between different parts of herself—a pessimistic part and a part that pushed her to achieve. Dr. Schwartz wondered if the voice that spoke so angrily of the pessimist was the part that pushed her to achieve.

Dr. Schwartz asked Diane to focus on the angry voice and ask it to stop interfering with negotiations with the pessimist. Amazingly, it agreed to "step back." Dr. Schwartz asked Diane how she now felt about the pessimist and she suddenly softened.

In a calm voice she said she understood that the pessimist tried to protect her and that she felt grateful. She could hear the pessimist's concerns about her life, just as she could hear the achiever's. But it turned out that neither the pessimist's need to protect nor the achiever's need for her to do more was her real Self.

In trying this "step back" approach with several other clients, Schwartz sometimes found he would have to ask two or three internal voices not to interfere or take over. Eventually the clients would end up in the calm state in which they felt compassion for their different parts but were not ruled by them.

Schwartz would then ask his client whom he was speaking with now. "That's not a part like those other voices are," the client typically answered. "That's more of who I really am. That's my Self."

Dr. Schwartz writes, "Once a person's parts learned to trust that they didn't have to protect so much and could allow the Self to lead, some degree of Self would be present for all their decisions and interactions."

Vocation—the way of being that we are called to—is not defined by our smaller desires and fears. It is the way we are when our protective, grasping parts relax and let the Self wander.

Because the Self is larger than the various thinking voices inside your head, you cannot predict where the Self will go. You can, on the other hand, see where it has gone in the past and so where it might tend to go in the future.

Each of us can be like the farmer and the horse. You do not build the corral where you want the horse to stay but where the horse naturally stays by itself. You don't build a life and then fit yourself into it but discern the tendencies of the Self and then build a life around it.

We discern the direction we veer toward when we look at the moments in our lives when we are free from the dictates of our inner protectors. At those times, in what direction do we head? In what ways do we act?

Sometimes we become so attached to our desires and fears that we begin to think they are the sum of who we are. It is like a kind of internal Stockholm syndrome where we come to feel affection for the emotions that have kidnapped our lives. When we consider putting our fears and desires aside, we feel we are being asked to be like the hole in the doughnut.

This is part of why it is hard for a Westerner to hear the Eastern philosophical tradition of no-self (notice I am using a lowercase *s* for self, and not an uppercase one). We think "no-self" means we will have to suppress our tendencies, sell all of our possessions, and find an orphanage to run like Mother Teresa.

We don't realize that if there is no self then there is also no other. There is no self to sacrifice and no other to sacrifice it for. There is just *this*—that which is happening right now, before the idea that there is some other experience to have or some part of this one to push away.

From this place—before we make self and other—we naturally follow a path that takes care of both ourselves and the world. This is the ground of our True Self, the place from which our vocation arises.

Sometimes, too, we get stuck in our need to understand. We agree that there is a more authentic path that lies beyond desires, aversions, and fears, and we plan to embark on it—once we have a better handle on it.

What if our problem is not too little understanding? What if our problem is too much understanding? What if our lives and Selves are bigger than our ability to understand? What if clinging to our understanding just means trying to force our lives into a smaller container? What if we never get to actually define our True Vocation at all because it is bigger than our understanding?

This possibility points to a different path: Letting go. Forgetting about corrals and fences. Opening the stable door and seeing where the horse goes. But to let the horse roam free, the farmer needs to feel safe. He has to find a way to trust the horse.

Learning to Let Go and Trust the Horse

According to Schwartz's Internal Family Systems theory, our mind naturally subdivides itself into a number of parts or subpersonalities. Each subpersonality contains its own talents, set of concerns, and approach to life. Every person also has a Self and, in our happiest and healthiest moments, the Self leads a person's internal system with input from the subpersonalities.

According to the theory, all our subpersonalities want something positive for us as individuals and try to influence our behavior to get what they want for us. However, sometimes, certain of our subpersonalities—called *protectors*—become extreme and take control away from the Self. In these times, we run our lives according to our desires and aversions rather than following our vocation.

The goal of Schwartz's "Self leadership" is to elevate the Self so it can lead the internal system while compassionately taking into account what the subpersonalities have to say. Part of the way our subpersonalities come to accept the Self's leadership is to come to trust that it will address their concerns.

So:

In the process of discerning your vocation, you may find that it is more of an intuitive process than a carefully strategized one, and you may not fully understand it. This can bring up a lot of fear.

As you try to learn about and discover your vocation and use your time in ways that feel more True for you, you will encounter resistant parts of yourself.

Voices may tell you that what you are doing is unrealistic or unproductive or stupid or won't lead anywhere or that you won't be able to support yourself and on and on. Those parts of you might keep trying to take over how you do things and make you scared of the prospect of "following your call."

When you encounter such a voice, wait until you have some quiet time, and then summon the voice back up. Assume that it is not trying to stop you from being who you are, it really just wants to help you. Once you calm it, it may have information for you about how to follow your calling while still taking care of yourself and keeping yourself safe.

Ask yourself, *Is this fear coming from the most centered, grounded, fullest part of my Self?* Try to locate that centered part of you inside yourself, even as you feel the subpersonality's fear.

Say to the part of you that is frightened that you are very interested in hearing its concerns, but could it please step aside and let your Self—that grounded part you just located—be in control for right now? You could even imagine your Self taking a seat.

Now, from a place of centeredness, as your Self, ask the frightened part what it is worried about or scared of. Listen to what it says and don't reject any of its points. What does it need in order to feel that you will be safe and taken care of? What irrational fears are there that you can reassure it about?

From your centered Self, suggest to the part that it is possible to find ways to take care of its concerns while moving forward into a way of life that is True for you.

Negotiate with that part of yourself. If it is worried about money, say, you might offer to spend some time making an emergency financial plan. The point is to find ways to actually listen to the frightened parts of yourself so that the Self can move forward.

Repeat this process with other parts of yourself that may have other concerns.

In this way, the frightened parts of yourself will see that you plan to take care, that you don't plan to be irresponsible but to be even more responsible, just in a different way. They can learn to trust the Self. Meanwhile, you may discover some of the ways you need your life to be in order for your entire personality to support your following your vocation. And you will learn to trust the horse.

How to Know Where Your Horse Goes by Itself

Who are you when you really let go? The question here is not who would you like to be, but who are you already and how do you act when you are relaxed and joyful and centered in Self?

When you think of the question this way, vocation isn't something big and far away and daunting and requiring major life changes. In fact, no matter how far we are from where we feel we want to be, each of us still experiences coming from the place of vocation—acting as we are called to—many times every day.

Let's call that experience *moment vocation*. Moment vocation appears every time we act in line with our True nature—in the absence of desire or fear or anger—through the day, regardless of our circumstances.

If you are unconsciously kind to someone at the Laundromat, if you quickly sketch a portrait, if you sing or act or do a math problem—whenever you are relaxed and express your Self for a second, this is moment vocation: being the way you are naturally called to be.

What most of us want most deeply—assuming our basic needs are met—is to live a kind of life where this spontaneous being our Selves is most able to happen again and again (for proof of this assertion, if you need it, look back at Part II: How to Want What You Really Want). One way to create the conditions that allow for that "spontaneous being" is to build our corrals exactly where we can really live. To shape our lives—not just our careers—after who we are really called to be.

To do that, it helps to discern our *life vocation*. Life vocation is what we might call the pattern that emerges if we connect all the moment vocation dots together. By the way, life vocation can change. It does not just come from within us, since it depends on the world and the world can change. Or we can change. It is not static. So it might even be better to call it our *life-for-now* vocation (how exciting that we may later in our lives get to have other adventures, too!).

Anyway, what we are going to do now is begin to discern our life vocation—or at least our life-for-now vocation—by looking at the pattern that arises when we string all the times of moment vocation together. What have we done, whom have we helped, and how have we been when we have followed our call in the past?

Where does our horse naturally end up when we open the stable door and let it roam free? When in our lives or even in our days have we found ourselves at the intersection of the world's greatest sorrows and our own greatest passions?

One way to structure this inquiry is to spend some time writing, though of course you could make images, too.

WHAT WORLD CONCERNS HAVE MOST MOVED YOU?

Take some time to write and write and write. Not for someone else to read. Just ask yourself, *What are the many things I care about in the world? What are things I have actually been moved to do something about?* What do you come back to again and again in your thoughts? In what way would you most like the world to heal? Then write as fast as you can. Tell yourself you won't stop until you've written four pages. Don't read it. The next day, do it again. Don't read it. Do it every day for a week or until you feel you are done. *When you are done, read over your pages and make a bullet-point list of the things that seem most important to you.*

WHAT PERSONAL PASSIONS HAVE MOST MOVED YOU?

The next thing is finding out how you can work in service of your concerns in a way that makes you happy. What are the skills you use and activities you do that make you lose yourself? Happiness is a kind of energy that, when we have it, can be used to help. It is not the destination, but it does provide some fuel. So, what has made you happy?

Do another writing exercise. Don't write what activities you *think* will make you happy. Research shows that we are very bad at predicting our future happiness. So, instead, write about the activities that have made you happy and fulfilled in the past. The kind of experiences that have made you forget yourself while you were doing them.

Do the same thing as you did for your concerns: write four pages as fast as you can each day for several days. *At the end, read through and make another list of important bullet points of the activities that make you happy.*

WHAT CIRCUMSTANCES SUPPORT YOUR PASSIONS AND CONCERNS?

Lastly, it might be good to be clear about what conditions you need to keep you comfortable in the world. Are you okay with not knowing how you will earn next month's rent until next month, or do you need a longer-term sense of security? Do you care if you are in hot weather or cold?

The previous exercise was about what you've done that makes you happy. This one is about what circumstances allow you to lose yourself. Are you okay being by yourself or do you need people around you? Again, try to stick with what you've experienced about yourself for sure rather than what you think might be true about you. Write your four pages a day about this for a week. *At the end, read through your pages and make a bullet-point list.*

Where Your Horse Will Rest

When you're done with all that, you have a list of the things you really care about, a list of what you like doing the most and the parts of yourself you like using, and finally a list of the conditions you need in your life.

The next step is beginning to envisage and slowly build a life situation that fits all these criteria. From all the things you've written, write out a way you want to be in the world that fits them all, where you use your passions to assist in solutions to the world's concerns in circumstances that support you.

Don't expect this exercise to identify a job. This is not a job book! Instead, what you will get is a set of criteria against which to measure the Truth of a job—and the rest of your life. It will help you discover how you want to use your life. Then you can choose to have a job (or not to have a job) that fits with that.

Remember, though, that you have to trust yourself throughout this process. It won't work if you write down things that you feel you *should* like or things that *should* make you happy. Don't write what you think your parents and teachers want. You don't have to show this to anyone. It is all private for you.

Personal Passion + World Concerns = Street Theater Against Fracking How One Person Found Her Calling-for-Now

Vocation, as we have said, is your inside job. Your outside job does not have to be the same as your inside job. Vocation is just about how you want to be in the world. The story of Monica Hunken, from Brooklyn, New York, is one example of how a person combined her passions and concerns to fulfill her vocation.

PASSION: THEATER

Where some kids' first passion is to draw or sing, back in middle school, mine was doing puppet shows and directing classroom skits and acting like a clown. I was way too shy to audition for a lead role, but a classmate got kicked out of the school play for stealing. I found myself playing the role of Nancy in Oliver!

I took it very seriously and said things like "I'm not Monica. I'm Nancy." One time, my mother was shouting at me in the car and I discovered that I could disappear into my character and feel protected. Later, I found I could use acting not to escape from myself but to really connect and be present, too. When I graduated from high school, I went to the experimental theater program at New York University.

My first day of school at NYU was September 11, 2001. I was walking across Washington Square Park and turned and saw the smoking tower. I didn't realize what was happening until I got to class and a girl shouted that the towers had collapsed. The class was divided into those who wanted to go off and drink and those who wanted to go give blood. I chose to follow a young Iranian woman who had been through this kind of thing before. She calmly led us to the hospital.

I hadn't yet become politicized as a white teenager from Northern California, and I didn't understand what could possibly cause people to take such violent action. I couldn't stop thinking, why would they do this to us? Why would this happen? What was the pain that could cause people to do this? And then George Bush said that the way we could all help was by shopping, and that struck me as completely wrong.

There were these communities around New York gathering in places like Union Square and Washington Square parks who wanted to heal the global rift. There were people standing on soapboxes and arguing about what to do about global injustice. I saw increasingly that our democracy was not working and that the Iraq War was about profit and war and energy and I started to teach myself street theater as a way to talk about these things.

I've discovered that my personal mission is to help people be fully expressed, so I teach performance workshops and perform myself. Citizenship is one form of expression. I became very involved in the performance aspects of a number of citizen groups working on civil liberties, free speech and the use of public space, including Reverend Billy's Stop Shopping Choir, Times Up! and Occupy Wall Street.

CONCERN: FRACKING

In 2008, director and activist Josh Fox showed me early clips of his film Gasland*—then called* Water on Fire*—and I became active in the anti-fracking movement. I began visiting community boards to advocate for a statewide ban on fracking and I helped organize fundraiser parties for the film.*

My father died when I was seven months old. He had been exposed to toxic chemicals as an engineer. He had burning sensations in his fingers and couldn't pick up me or my sister and he didn't realize it but he had cancer. Fracking, the "hydraulic fracturing" method of drilling for natural gas, can poison people's drinking water and potentially bring radioactive gas into their homes without their knowing. What happened to my father—being unwittingly poisoned—was similar. So fracking—for personal as well as many political reasons—particularly concerns me.

Before long, I learned of a plan to bring a fracked natural gas pipeline—operated by a Texas-based company called Spectra and ConEd—under a public park and right past a chil-

dren's playground on the Hudson River and into Manhattan. A bunch of us from the Environmental Solidarity Working Group from Occupy got together and we began to organize against the Spectra pipeline along with Reverend Billy and the Sane Energy Project.

From that point, there were pretty much always demonstrators outside the Spectra construction site. We called the group Occupy the Pipeline. We shut down construction several times through acts of civil disobedience. I participated in a lot of ways, including using my theatrical skills in our many actions, rallies and trainings, which often incorporated theater, dance and puppets.

As one example, the governor of New York, Andrew Cuomo, had convened a conference of Democrats at the Sheraton Hotel in the city as part of his exploration of a potential presidential bid. A few of us snuck in. Some of our team hung a huge banner from a window, others participated in a rally outside, and some of us—including me—prepared for a disruptive action inside the room where Cuomo was speaking.

When the subject of fracking came up, my friend, who is a magician, jumped up and shouted that he would now demonstrate his "magic frack act." Once he had everyone's attention, he took a clear glass of water and turned it green.

"As Jesus turned water into wine," he shouted, "now through the miracle of hydraulic fracturing and Halliburton, Governor Cuomo will turn New York's clean water into brine."

I then stepped up, gulped down the green water, and proceeded to act as though I had convulsions. When the act was over, we were dragged out by security guards to major applause—I'd like to think it was for our act! More important, Governor Cuomo was forced to discuss fracking at the summit and it was covered by the media.

In December 2014, Governor Cuomo finally banned fracking in the state of New York, citing health risks. Our work in Occupy the Pipeline was, of course, just one small part. So many people in so many groups have worked so hard to stop fracking and the building of the fracking infrastructure in New York. There is still work to do. But knowing we were part of this accomplishment sure feels good.

The Vocation of No Vocation

Some of us have no idea what we want to be or are quite content with our lives and circumstances. We don't have a drive to change our conditions, or, even after doing the exercises above, we still feel confused about our destination. Or we don't have a skill or talent we particularly want to engage; we are happy with who we are. Still others of us have too many responsibilities to change our situation for now.

That is okay. A fancy job title or career aspiration does not a vocation make.

Suppose, for example, you head a huge human rights organization. Every day your organization feeds hundreds of thousands of people. But you personally are constantly on your phone and sending e-mails and your friends and family aren't happy with your level of distraction. Your stress levels make you unkind to waiters and taxi drivers. You have no compassion for your staff. You may be the head of a big philanthropic organization, but in those moments, are you really following your vocation?

On the other hand, suppose you are a cafeteria worker at a small university. Every day you come out from behind your counter to clear dishes off the tables. It may not seem like a great job description, but what if clearing dishes is your excuse to sit and joke with the students? Suppose sometimes a student sits alone and you find yourself asking after her well-being? Maybe you think she needs a little support and you invite her home for a dinner. You are a lunch worker, but in those moments, aren't you really following your vocation at least as much as the unhappy NGO head?

A butcher who lived in the time of Buddha was on a spiritual quest to find his True Self. Finding peace, he had been taught, required following a nonviolent path, and the first Buddhist precept is "Do not kill." How could he find peace as a butcher? he wondered. On the other hand, he had a wife and small children and knew no other skill. He could be a butcher or he could let his children starve.

One day, our butcher sought out Shariputra, a monk who had been one of Buddha's top disciples. The butcher said to Shariputra, "I kill cattle all day, so I'm doing very bad things. I will never find my True Self."

Shariputra said, "What is the essence of what kills the cattle? What are you?"

These were exactly the questions at the root of the butcher's spiritual quest. The one he thought being a butcher might stop him from answering.

The butcher said, "I don't know."

Shariputra said, "Just keep that not knowing. Pay deep attention in every moment. Don't attach to your good and bad thoughts about your situation. Just watch them come and go."

So instead of wallowing in his bad feelings about his work, the butcher stayed very curious. *What is this? What is happening now? What is the fundamental essence of what is happening now? What am I?* All sorts of thoughts came into his mind, but instead of believing what the thoughts said, he asked the questions about the thoughts, too. *What is the nature of these thoughts?*

This constant inquiry meant he never trusted his thinking opinions but they instead forced him to watch carefully and act meticulously. Because his opinions and concerns and worries about his work no longer stood in his way, he had a new clarity and he was often able to be exactly as he was called to be. Not *understanding* his True Self, he was able to *be* his True Self.

So when hungry people came with no money to pay, he could see their suffering and give them food. When a cow had a young calf, he could see that the calf needed its mother and let the cow live longer. He became a different kind of butcher. In fact, he became a sort of saintly butcher. At all times in all places a simple question rose in his mind: *Given my situation, how can I help?*

Following our calling does not depend on our circumstances. It is not the job title alone that matters but the attitude we take toward doing the work. Why are you alive? Why do you do your job at all? There is only one calling: *How can I be of help? How can I live "not just for me"?*

We can use our conditions to help others. Use our circumstances. We don't have to wait until we are in another place. This is our all-the-time job. We mustn't be distracted from things as they are by thoughts of how we wish they were.

Whatever your job title, at all times, in all places, simply ask, "How can I help?" and, no matter what your circumstance, your vocation will reveal itself before you, one moment at a time, like the direction of a walking path in a thick fog. This is where you continually respond with kindness or compassion at all times and all places. This is the vocation of no vocation, and it is our highest calling.

Three More Exercises to Help Center You in Your Calling

1. All day long, just pay deep attention to what is happening in the moment and ask, "How can I help? How can I help?" Don't worry if nothing happens. Don't worry if you get distracted. Just come back to the moment and the inquiry. How does your behavior change?

2. Schedule some mini-sabbaticals. We are so tyrannized by the clock and the calendar that we march to their rhythms instead of our own. Make no plans for a dedicated hour, or day, or week. What do you find yourself doing? How do you find yourself acting? Repeat and repeat.

3. Watch for the small heroic actions of others. All day long, look for ways of acting that you admire and just notice them. See if you can find a way to copy the action as soon as possible. If you like that someone flatters a stranger, then do it yourself as soon as possible. What do you learn about yourself?

27

THE LIFE OF YOUR TIME

Five Hundred and Fifty Thousand Hours

Five hundred and fifty thousand hours is about the amount of time the average twenty-one-year-old recent college graduate has left to live. It's the time those of us lucky enough to get a college education have to choose what to do with once the schooling is over. No more homework and here is a gift card with 550,000 hours on it. Go!

The thing is, between sleeping, eating, washing, and exercising, most of us spend about half those hours maintaining our bodies. That leaves us with 275,000 hours total, or about 3,275 84-hour weeks to spend on what is important to us. Of course, each of us, regardless of our age, gets those 84 hours each week. It's just that those of us who are older than twenty-one have substantially fewer than 3,275 of those weeks left.

This is the life of your time. It is pretty precious, when you look at it that way, isn't it?

Knowing What to Do with Your 3,265 Weeks

Sometimes we become so caught up in the "realities" of life that we forget what we hoped to do with it. Return yourself to these questions:

What did you want to be when you grew up?

What have you always wanted to do that you haven't done yet?

What have you done that you are really proud of?

What do you want to learn?

How could you be more self-reliant?

Where have you always wanted to go?

How to Be a Productive Member of Society

For many years, in our culture, the standard life approach has assumed that to be a productive member of society means to participate in the market economy, to use forty or more of our weekly eighty-four hours in the execution of a job. That is why when colleges talk in their promotional literature about helping young people to become productive members of society, they pretty much mean to get a job, any job.

As the standard life approach goes: Everything everyone needs is produced by the market economy. Economic growth means that more people will get more of what they need. Therefore, the primary goal of society should be economic growth, and to be a productive member of that society is to assist in that growth by having a job.

This is why, according to the standard life approach, those of us who don't have jobs are freeloaders. Not just because we potentially live on other people's money but because not having a job means we don't contribute and are not productive members of society.

Many of us understand that things have become much more complicated than that. Nowhere near all the societal good and quite a bit of its bad actually comes from a continued societal emphasis on growth as opposed to, say, sustainability (or better yet, regeneration). Partly for this reason, and for lots of others, too, many of us worry that our having a job is the opposite of contributing. Some of us are sadly stuck in jobs and working for organizations we feel degrade the world.

On my Facebook page, I asked my fans and friends these questions: What does it mean to you to be a productive or contributing member of society? What do you need to do to feel you are contributing in this world of today?

Here is some of what they said:

> *Working close to home, supporting those around my neighborhood.*
>
> —Sharley Babul, Airdrie, Alberta

> *Promoting acceptance of sustainable lifestyles.*
>
> —Tanya Seaman, Philadelphia, Pennsylvania

> *Living my life without reinforcing a system that puts some ahead of others.*
>
> —Marlin Ballard Jr., Baltimore, Maryland

> *Helping lots of people interrupt the economy-as-usual.*
>
> —Billy Talen, Brooklyn, New York

> *Sharing that which I have learned—to teach.*
>
> —John Lundin, Minca, Colombia

The problem is that, for so many of us, our jobs contain none of the things on the list of what we feel makes us productive members of society. In fact, in a major turnaround, while some people forgo having a job or get paid less in order to be a truly productive member of society, those of us earning high salaries in high-paying jobs can't help feeling like we are now the freeloaders.

The central life-planning question for so many of us used to be, *How do I find a job in order to be a productive member of society?* For so many lifequesters, that question has now become *How, in spite of the fact that I have a job, do I find a way to use my weekly eighty-four hours to be a productive member of society? How, in spite of the fact that I have a job, do I live my eighty-four hours in line with my calling?*

How to Be a Productive Member of Your Own Life

If you have taken anything in this book to heart, you embrace the understanding that the lifequester's path is both good for the lifequester and good for the world. If you sacrifice what is truly good for you, you help the world less. If you sacrifice your own particular brand of doing what is good for the world, you help yourself less.

That means that to be a productive member of your own life requires finding a balance between using your eighty-four hours to make yourself happy and the world happy. It means giving energy to both your inside job and your outside job. So the question is, assuming you are a lifequester looking for your own version of the Good Life, of those eighty-four hours each week, how many would you like to spend doing each of the following?

Helping the world according to your calling

Earning money

Hanging with friends, family, and people you love

Lazing around, reading or watching media

Playing sports or games or other fun stuff

Engaging as a citizen in your local, state, and national democracy

Learning new ideas and skills

Exploring your spiritual life

Making art and other cool stuff

Spending time in nature

Enjoying the culture

Working for your local community

Doing errands, chores, and life maintenance

Goofing off (a truly spiritual act—seriously!)

Do It!

Your Ideal Eighty-Four Hours

How do you want to use your time? How much of it to help the world? How much of it to simply enjoy? How much of it to grow? How much of it to earn money? Look at the list above and add to it if you want. These are the ways you can spend your time.

Now assume, for now, that you don't actually *have* to earn money. We will deal with so-called realities later. In an ideal week, how would you break up your eighty-four hours and what specific activities would you choose that fit into the categories above?

Write down your answers and keep the list. This is your vision for an ideal week.

Earn to Live or Live to Earn?

Okay, so maybe your job isn't everything you want it to be. Maybe you don't get to use your time in line with your calling or even in line with the other ways you just wrote down you want to spend your time. But go to the job, work hard, earn a pile of money, and then you can use all your free time exactly the way you want and have the money to do it. This is the standard answer to complaints of meaninglessness in our work.

Here is the problem: the idea is that you trade a certain amount of time to be used in ways you don't want in return for making the rest of your time much better, right? But let's examine how well that works. To get to where we want to go, it is good to understand where we already are (and don't worry, we will turn to fixes soon).

So:

Let's think of an imaginary young woman named Jess who has taken the standard life approach to chasing after career. She just graduated from college. She has moved far from her home in California to a town that is just okay. She loves to surf but now she is landlocked. But she believes these are sacrifices she has to make because she has been fortunate to land a well-paying job.

Let's assume Jess spends 45 hours a week at work. That leaves her with 39 hours out of her weekly 84. But Jess also dedicates another weekly 5 hours to commuting and another 5 to work-related chores like taking her work clothes to the dry cleaner and answering e-mails.

For the sake of argument, let's assume that Jess needs another half hour each workday just to recover from her job, 30 minutes during which she is too tired to do anything else. Let's assume, too, that Jess spends another 30 minutes each day doing the chores she needs to do to run her life—everything from paying bills to grocery shopping.

All told, this leaves Jess with 24 hours of unstructured time each week to do with as she chooses. Looking at the list of the ways most of us want to spend our time above, that means Jess has 24 hours each week to spend doing everything from being creative to tossing a Frisbee to goofing off to touching base with her calling.

Since 80 percent of workers are dissatisfied with their jobs (45 hours out of each weekly 84), the big prize is supposed to be that her 24 free hours will totally rock and so will her vacations. The sad thing is that, because she moved for her job, Jess no longer gets to surf in her off hours. Nor is she near the people she loves. She assumes that she will get to surf and see her loved ones during vacations. The problem is that, on average, Americans take only half of their allowed vacation time.

Is the payoff for the "great career" big enough? Because don't forget, Jess has only 3,275 84-hour weeks on her gift card before it is all used up. How many is she really willing to use this way? With a limited number of 84-hour weeks left, how many of them are any of us really willing to use this way?

Why Nice People Who Work for Corporations Have to Do Bad Things

In his book *Moral Mazes: The World of Corporate Managers,* Robert Jackall reported on his several-year-long study of ethical behavior within large organizations. One of his main findings was the grueling long-term effect of the fact that power concentrates at the top of an organization while the responsibility for making decisions and profits concentrates at the bottom.

This means that even if the CEO would personally agree with the ethics of your making a "right and moral" decision, he will never know the details of your daily decisions. Instead, he will fire or promote you according to whether your department achieved or didn't achieve his goals for it. People who resist this fact tend to get drummed out of their organizations, Jackall found.

Living in this kind of environment over time causes people to spend less energy considering the morality of their decisions and more time thinking about what they have to do to survive and succeed. One corporate vice president told Jackall, "What is right in the corporation is not what is right in a man's home or in his church. *What is right in the corporation is what the guy above you wants from you.* That's what morality is in the corporation."

In fact, as part of Jackall's research, he presented several case studies that included moral quandaries to experienced middle managers. He asked them to assess the performance of the key players in the case studies. The middle managers had become so inured to the morality of corporate decision making that they completely overlooked the fact that the cases even had an ethical dimension.

Instead, they assessed the characters in the case studies according to their compliance with a generic set of bureaucratic rules that they each gave in varying forms to Jackall:

> (1) You never go around your boss. (2) You tell your boss what he wants to hear, even when your boss claims that he wants dissenting views. (3) If your boss wants something dropped, you drop it. (4) You are sensitive to your boss's wishes so that you anticipate what he wants; you don't force him, in other words, to act as boss. (5) Your job is not to report something that your boss does not want reported, but rather to cover it up. You do what your job requires and you keep your mouth shut.

How Much Do You Really Pay for Your Money?

Couples in which one partner spends ten or more hours more than average at work are more than twice as likely to divorce.

Twenty-five percent of employees say work is the main source of stress in their lives.

Even on vacation, 25 percent of us check in with work at least hourly by e-mail and phone.

Two in five workers say they've gained weight at their jobs due to sedentary all-day desk lifestyle, stress eating, and food found at the office.

It's Saturday. Our imaginary friend Jess starts off the day by taking the work clothes she wouldn't normally wear to the cleaner and the car she wouldn't otherwise drive to get washed. She spends an hour at the grocery store buying frozen food she wouldn't have dreamed of eating a year ago. She no longer has time to cook or go to the farmers' market. Feeling like she isn't dressed as well as her colleagues, she goes to the mall. She didn't have time to plan anything else to do anyway.

In other words, although she is not at work, Jess spends weekend time and personal money getting herself ready to go back to work. Clearly, she is spending more than her contracted forty-five hours on her work and effectively getting paid less than her pay stub says.

Vicki Robin and Joe Dominguez, coauthors of *Your Money or Your Life: Nine Steps to Transforming Your Relationship with Money and Achieving Financial Independence,* suggest that we ask ourselves exactly what we are paying for our money. How much total life energy—not just time in the workplace—are we really trading for our paychecks? And how much of that money do we really get to spend on ourselves and not on costs related to our jobs?

To figure this out in your own case, start by calculating how much time and money you spend maintaining your job. For example, determine how much time and how much money you spend each week:

At your workplace.

Answering e-mails or doing other work outside your workplace.

Commuting back and forth to work.

Buying and servicing clothes you wouldn't normally wear.

Eating food in places you wouldn't normally go for prices you wouldn't normally pay.

Paying extra rent to live where your job is.

Decompressing from the job. (You wouldn't have to do it if you didn't have the job.)

Indulging in escape entertainment because you have no energy to do anything else.

Going on vacation or home for Christmas. (Would you have to go on vacation if the job didn't have you living somewhere you didn't want to?)

Sitting in the therapist's or life coach's office dealing with job-related stress.

Paying babysitters you wouldn't need without the job.

Any other expense or activity caused by your job.

From the list, add together all the time you spend related to your job, from being at your workplace to answering e-mails at home to business travel to job-related errands. Next, add up all the work-related expenses and subtract them from your weekly after-tax income to get your total real weekly salary. Now divide your real weekly wage by the total job-related time you spend to get your real hourly wage.

So, for example, let's say you earn $1,000 a week for a contracted 40-hour workweek. Ostensibly, you earn $25 an hour. But you pay a total of around 25 percent in taxes—which means you actually take home $800 a week—and you spend $300 on job-related costs. Also, though you are contracted to work 40 hours a week at your job, you actually spend 50 hours there and spend an additional 30 hours on job-caused tasks.

Your actual hourly wage is $500/80 = $6.25 an hour.

This means you are paying nearly ten minutes of your life for every dollar you earn.

Now a worker like our friend Jess might say, *Well, that is the cost of my lifestyle. I like to be able to drink my four-dollar cappuccino every morning and have my twenty-dollar mani-pedi once a week.* When you realize that you are trading, say, ten minutes of your life for every dollar, that means that Jess is trading more than three hours of life each week for those daily cappuccinos and another three hours for the mani-pedis.

Looking at the list of what you really want to do with your life, would you rather spend those six hours on coffee and nail polish or on free time to do what really matters to you? Suppose Jess had never moved away from the ocean for her well-paying job. Yes, she might earn less. But couldn't she then reduce some of her expenses by making really good coffee at home and putting nail polish on herself, reclaim those six hours of life energy, and use them to go surfing?

I'm not saying that there are no realities in life, that we are free to do whatever we want. The point is to ask how much better our lives might be if we chose our realities—built our corrals—so that we could use our time in a way more closely aligned to our values.

28

THRIVING IN YOUR CORRAL

Don't Forget That a Quest Is, Well, a Quest

You know your vocation and you know how you want to spend your weekly eighty-four hours, or at least you have an idea. The question is, how do you create the life and work circumstances that fit? How do you get there? How do you build your corral right where you live so you don't feel the need to leave?

The truth is, there is no connect-the-dots answer. Our whole culture is in the wilderness, and lifequesters are leaders in the search to find a way out. Societally, we are in the transition between what used to work in the past and what will hopefully work in the future. No new cultural story has yet been set down to replace the old.

That is why finding the authentic Good Life is a quest. If you take to heart the suggestions in Part Four: Finding Your People, it won't be long before you find support from people who are in the same boat. But ultimately you must feel your way along the path of Truth, whatever that means for you. That might seem scary—that's in the nature of a quest, too—but it's also exciting.

Bootstrapping Your Way to Where You Want to Be

Bootstrapping is a term used in entrepreneurial business that more or less means using your current assets to move in successive, small steps toward your final destination. Your new, stronger position then provides a new, perhaps unforeseen opportunity to take another step in the direction of your destination. One step at a time, you pull yourself along by your bootstraps.

One advantage of bootstrapping is that you build and make changes without having to find a big investor or lender. You can do what feels right to you and is authentic to what you are trying to do without needing to convince anyone else to come along for the ride. Also, because you are taking small steps, you can anticipate the gains and risks.

Bootstrapping our way to the Good Life has the same advantages. You take good stock of where you are and then decide what small changes you can easily make. The risks are small, the advantages are predictable, and you don't have to make huge, disruptive changes—at least not before you have determined that disruptive change is the only way to go. You can also decide if the direction you are heading feels right and change course easily.

Basically, bootstrapping is a nonviolent process of giving more energy to what is True for you and giving less energy to what is not. It is an over-and-over cyclical process that leads us ideally to the place where only tiny adjustments keep us on our life course or to where we see we need to leapfrog and make a big change to get to where we want to go.

How You Spend Your Life Energy

If you ask someone for directions to get somewhere, the first thing they will ask you is "From where?" Similarly, in order to know how to get somewhere in your life, you need to know where you are now.

Thus, to know what we want to give more and less energy to,

we need to know what we spend our life energy on now. By understanding how much each thing offers pleasure, meaning, service and alignment with your values and calling, you can decide whether to give it more or less energy.

For our purposes, life energy consists of (1) time and (2) the money for which we have traded time. So, in order to see how we spend our life energy in any given week, we must record what we pay for and how we spend our time in that week.

1. Record: For one week, write down everything you spend money or time on in dollars or hours both during your working and nonworking hours. You can do this on paper or you can use any of the many financial and time logging apps (if you use a debit card, all spending will be automatically entered if you connect your financial app to your bank account).

2. Categorize: At the end of the week, categorize your life energy spending in a way that makes sense (rent, food eaten out, groceries, transportation, etc.). Add together the number of dollars or hours spent in each category. Don't forget to add any electronic debits that you did not write down. Also, look over your bank statement for any monthly, quarterly, or annual expenses (car payments, bank fees, etc.) and add the weekly equivalent of those expenses to your list. Add to the list anything else you might also spend life energy on in the average week that you know is missing.

3. Equalize: If you figured out how much your money cost you in life energy on page 382, convert your monetary spending into hours. You now have a list of everything you spent life energy on in your week in hours.

4. Fantasize: Look at your list of how you would ideally spend your eighty-four hours from page 376 and add any activities you *wish* you had spent time on this week (prospective life energy spending) to the bottom of your categorized life energy list.

5. Analyze: Now go through your list and give each category of

Category	Dollars Spent	Hours Spent	Life Energy Equiv. (hours)	Value	Align-ment	Desir-ability	More/ Less?
Weekly Cash Spending							
Rent	300		48.0	3	3	1	-
Commuting	40		6.4	3	3	1	--
Heat/electric	80		12.8	3	3	3	-
Car payment	100		16.0	3	2	1	-
Cable/internet	80		12.8	2	2	2	0
Groceries	80		12.8	4	4	4	+
Work lunch	50		8.0	3	3	3	-
Dry cleaning	20		3.2	1	1	1	-
Entertainment	20		3.2	4	3	3	+
Coffee	20		3.2	1	1	3	0
Nights out	40		6.4	4	4	4	0
Travel fund	0		0.0	5	5	5	++
Surf lessons	0		0.0	5	5	5	++
Etc.	--		--	--	--	--	--
Weekly Work-Related Time Spent							
Client contact		20	20	5	5	5	++
Travel to client		10	10	2	2	1	-
Client reports		10	10	3	3	3	0
Team meet'gs		5	5	2	3	2	-
Admin		5	5	2	2	2	0
Further educ.		0	0	5	5	5	++
Mentoring		0	0	5	5	5	++
Etc.		--	--	--	--	--	--
Weekly Non-Work-Related Time Spent							
Sleeping		49	49	4	4	4	+
Eating		7	7	4	4	4	+
Self-care		5	5	3	3	3	0
TV		14	14	2	2	2	-
Work errands		10	10	2	2	2	-
Socializing		5	5	5	5	5	+
Surfing		0	0	5	5	5	++
Time with fam.		0	0	5	5	5	+
Cooking		0	0	5	5	5	++
Etc.		--	--	--	--	--	--

real or prospective life energy spending a value from 0 to 5 in each of the following areas. In the case of activities, this is not an assessment of whether they earned you money or not but an assessment of whether the activities are in line with the person you are called to be:

> V(alue) = the amount of pleasure, meaning, or satisfaction received for life energy spent

> A(lignment) = the degree to which life energy spent aligns with values, purpose, and calling

> D(esirability) = the chance that you would use life energy on it if you didn't have to earn money by working (a suit might get a 0, for example, if you normally wear jeans)

6. Finalize: You now have a list of how you spent your life energy. This is an excellent place from which to start bootstrapping your way into the approach to life that you are called to. The more you repeat this exercise, the more accurate your understanding will be. Now that you know where you are, you can figure out how to get to where you want to be.

How You Would *Like* to Spend Your Life Energy

Now that you understand where you are, you can figure out what direction you want to move in. The amazing thing about the bootstrapping approach is that you don't have to know exactly where you're going, you just have to know what you do and don't value about how you spend your life energy now.

So:

What you now have is a categorized list of how you spend your life energy at home and at work, how you wish you spent your life energy, and how much value you get from each category. Our next step is to go through the list carefully, one category at a time.

Look at each category. Don't ask yourself if you *have* to spend that life energy on a thing. Ask if, in your Self, spending life energy on a thing is in line with who you are. Next to each life energy expenditure, put a 0 if you are spending the amount of the life energy you want on that category, put a + or a ++ if you wish you spent more or a lot more energy, or put a – or a – – if you wish you spent less or a lot less energy.

Think deeply. For example, you may think, *I have to pay my child's private school tuition.* And it may, of course, be true. But do you happen to be the kind of person who wishes she had the time to homeschool her child? Or who believes in public education but lives somewhere because of her job where public school doesn't feel like an option? Put a – by tuition, add a category for tutoring your kid, and put a + by that. (Of course, if you wish you were paying tuition for your child, put that in with a +.)

If you are spending a lot of time in traffic on your way to work and you really dislike it, put a – or a – – by that item. If you really wish you were practicing guitar much more or singing in a choir, put a ++ by the corresponding category.

In terms of work time, you will find that since you have broken your job down into tasks, there are things you really like and would like to do more of and things you really dislike. If you hate the weekly staff meeting and find it of terribly low value, put a – – . If you love mentoring a younger employee, put a ++.

When you are done, what you will have is a guide to exactly how you would like to change the way you spend your life energy. It is like a list of directions that say, "This way to the corral you want to build."

Now, all you have to do is choose a couple of things you want to do differently, and add a little more time, say, or subtract a little more money from a category according to your pluses and your minuses. Look at your work calendar. Are there any meetings you put a – next to that you could shorten? Would this give you a little more time to mentor the young employee?

If you repeat this whole exercise over and over again, you will find yourself marching along the path to where you want to be. That is bootstrapping.

We are next going to discuss three approaches to adjusting your work life to bring it in line with your vocation and what you have learned from your bootstrap list, in order of least to most disruptive (and probably least to most life changing):

Thriving in your prefabricated corral: Adapting your current work situation to bring it more in line with your living vocation.

Choosing a different prefabricated corral: Finding a whole new job that is more in line with your living vocation.

Building a custom corral exactly where you live: Learning to bend your job choice around your Good Life instead of bending everything about your life around your job. We will devote a whole chapter (chapter 29) to this.

Thriving in Your Present Corral

Remember the butcher who couldn't change his job but was able to find his calling anyway? His corral was built where he didn't want to be and the fences were too high to jump. That is what life is like for many of us. We have so many commitments that we can't jump the fence and go somewhere else.

But what our butcher did, since he couldn't move his corral, was change what it was like inside the corral. He learned to thrive in the conditions he had. What you have that he didn't is the bootstrap list that you have developed with all sorts of potential changes—large and small—that will help you thrive without changing your corral.

Look at your list of work activities. Play with your calendar and your daily routine to give more time to the items with a + and less time to items with a –. Think of creative ways to make your day more meaningful by reducing the time expenditures that are meaningless to you and increasing the ones in line with your calling.

If you find a way to move one hour of work with a -- by it out of your schedule and one hour of work with a + by it in, that is big progress. Do that ten times—for ten hours in total—and you will have made your work corral substantially more livable. Think about what tasks you can:

Say no to.

Cancel.

Delegate.

Negotiate out of or into your job description.

Do differently to make them more meaningful.

Trade with others for tasks you like.

Crafting Meaning Into Your Job

Back in 2001, Amy Wrzesniewski and Jane E. Dutton, then at New York University and the University of Michigan, respectively, published a paper in the *Academy of Management Review* about what they called *job crafting*. Job crafting is when an employee slowly but surely recrafts the tasks, relationships, and perception of his job to make it more meaningful.

Wrzesniewski and Dutton identified three types of job crafting that can help you add + tasks to your job and take – tasks out, so that you can thrive in your current corral.

1. **Task crafting** consists of adding or dropping tasks, adjusting the time or effort spent on tasks, and redesigning aspects of tasks. For

example, suppose you work in a hospital fund-raising department writing promotional copy. Spending more time with children is on your life-energy wish list. You might ask your boss if you could spend some time in the children's ward making videos for the website.

2. **Relationship crafting** consists of creating and/or sustaining worthwhile relationships with others at work, spending more time with preferred individuals, and reducing or completely avoiding contact with others. Suppose you have a passion for sustainability, for example, but you work in public relations at your company. If you got to spend time with the operations manager, you could help come up with sustainable energy or purchasing plans and then help promote the changes.

3. **Cognitive crafting** consists of employees' efforts to perceive and interpret their tasks, relationships, or job as a whole in ways that change the significance of their work. In one of Wrzesniewski and Dutton's studies, one group of hospital cleaners experienced their work as boring while another group experienced it as engaging and meaningful. The second group achieved this meaningfulness by changing the way they thought about their jobs.

They interacted more with doctors, nurses, and patients, and took it upon themselves to make everyone's experience better. Generally, they saw their work in its larger context. They didn't just take out the trash and dirty laundry. They were an important part of the patients' well-being and the hospital's smooth running.

Indeed, hospital cleaners who saw their work as their calling were happier than doctors who did not.

Social intrapreneur, n.

1. Someone who works inside major corporations or organizations to develop and promote practical solutions to social or environmental challenges where progress is currently stalled by market failures.

2. Someone who applies the principles of social entrepreneurship inside an organization.

3. One characterized by an "insider-outsider" mindset and approach.

—From SustainAbility's report "The Social Intrapreneur"

In their book *Making Good: Finding Meaning, Money, and Community in a Changing World,* authors and change agents Billy Parish and Dev Aujla quote the novel concept of the intrapreneur. Intrapreneurs, they say, are people who work from the inside of an organization in order to promote outside ideas. Because they understand the language, culture, and power dynamics of the organization, intrapreneurs are in a great position to have a major impact.

In their book, Parish and Aujla give a road map for becoming an intrapreneur:

Reach out: Become an insider/outsider by learning more about the outside ideas that you want to bring into your company. If you are interested in combating institutional racism in your organization, for example, go to conferences and take meetings with people who are making progress on that issue in their own organizations.

Translate: Figure out how the ideas you are learning about translate into the culture, language, and priorities of your organization. Does your company want to be a hip social leader? Does it want to attract the best talent? Does its marketing department want to reach new communities? How are you able to frame what you have learned about combating institutional racism in ways that play to what your organization's leaders perceive as its self-interest?

Find powerful allies: Map out the people who are the levers for change in your company. Will the director of human relations respond to a direct appeal? Will the head of public relations or marketing help make your case?

Take a baby step: Although you may ultimately hope to change your organization's entire policy and culture with regard to hiring and racism, think of a pilot project that could help convince your allies to expand to a larger project and then a new policy.

Marshal your proof: With these pieces in place, you now have ammunition and allies to campaign for bigger change. What is your strategy?

Crafting Meaninglessness Out of Your Job

Here are some examples of typical ways your time might be wasted at work:

> It takes you an hour to get to and from a job you could just as easily do from home. On your bootstrap list, there is a double minus by commuting.
>
> Your boss walks past your desk on the way to her meeting and, on a whim, asks you to come to the meeting. You neither learn nor contribute anything important. Useless meetings have a double minus by them.
>
> The office gossip stops by your desk. A colleague has an emotional crisis you can't get out of discussing. An intern is assigned to sit with you for the day and asks endless questions. Three hours after you got to the office, you've done forty-five minutes work. Worthless interruptions have a double minus by them.

In his book *The 4-Hour Workweek,* Tim Ferris talks about how getting your boss to agree to let you work from home can get rid of whole swaths of meaningless work distractions. You could do more of what the boss wants you to do in half the time or less. This leaves you with more time to tutor kids, go on a beach cleanup, take a hula hoop class, go on a climate march, or do any of the other things with pluses by them on your bootstrapping list.

In order to get your boss to agree to let you regularly work one, two, or more days from home, Tim suggests a process that goes like this:

Prove your worth: Spend some time making your work (not your physical presence) indispensable. Make it so your boss would hate it if you quit. Use the job-crafting techniques above to find and fulfill a role that no one else easily could.

Demonstrate increased output: Call in sick for two consecutive days—not including Monday or Friday, so it doesn't look like you wanted a long weekend. Double your regular work output on those days, perhaps using software that allows you to access your desktop computer remotely. Use e-mail or other means to leave a trail that documents this increased output.

Detail the business benefits: Make a bullet-point list of the benefits to the boss of the two days you spent at home. Show how the boss quantifiably got much more of what he wanted by your being at home. Also show how you ameliorated any disadvantages.

Propose a revocable time period: The week after your sick days, go to your boss to bounce an idea off him. "You know, our workload is really high at the moment and I kept track of my time last week. I noticed I accomplished twice as much. I was thinking if I worked at home one day a week for the next couple of weeks, that would really help us. If I'm needed here I'll just come in." Make the one day the most productive ever.

Expand the time period: Present your boss with a bullet-point list of the benefits. Show how you used Skype, phone, and e-mail to fulfill all your communications responsibilities. Suggest that you could really make a big splash if you could concentrate on the important tasks at home for a two-week trial period of whatever number of days you feel you can get away with. Keep documenting the increase in output.

Choosing a New Corral

Sometimes, though, you simply cannot reconfigure your existing corral so that you can authentically live a Good Life within it. You started out working in data management, say, because you were good at it, but you realize you are way more concerned with homeless youth. There are so many double pluses and double minuses on your bootstrap list that stay exactly the same week after week.

Remember the butcher? His story does not mean that we are supposed to stay in a situation that doesn't work. The butcher had to. He was a member of the lowest caste and could never change. He had to thrive where he was.

You, on the other hand, do not. Yes, you have some "realities," but is there another corral—another job—where you could reduce the number of immovable pluses and minuses on your bootstrap list? Finding a new job or a new career is something that is so well discussed elsewhere that we will not go into it here. But here are two resources that look at the job and career search from a perspective close to the lifequester's—with vocation and calling and moving toward the Good Life in mind:

Making Good: Finding Meaning, Money and Community in a Changing World, by Billy Parish and Dev Aujla: A great book about launching a new career or adapting an existing one to help you do good in the world.

Idealist.org: The top job search engine for people wanting to work in the social change movement.

29

BUILDING A CORRAL WHERE YOU LIVE

When Even a Meaningful Job Doesn't Work

Remember Jess, our surfer friend who got the new job far from the beach and discovered she was really getting $6.25 per hour of life energy when, at first, she thought she was getting $25.00? Let's assume it's a few years later. She finally found she couldn't keep her horse in that corral and she found another corral, one more in line with her calling. Let's say she now works with children through an educational nonprofit.

But the math, she finds, is pretty much the same. She is supposed to be working 40 hours but she really works 45. In her off hours, she spends the same amount of time servicing her job, and she also has her household chores, so she really has only 24 free hours. And of course she took a pay cut, so that means she is now earning a time-energy-adjusted wage of about $5.00 an hour.

All of that could be okay. But Jess is keeping her bootstrapping list going and she sees that for all of that hard work and continued job crafting, she is still spending so much time traveling between clients, writing reports on clients, and in office meetings that she is actually spending only 15 hours in contact with her clients. She spends only 15 hours a week with the kids.

In other words, she supposedly has the meaningful job, but is she doing meaningful tasks? Meanwhile, all the things on her bootstrapping list she would do in her off hours are just staying there, not changing. Especially the surfing.

First, Jess tried the standard life approach, which is to figure out the most palatable way to get the biggest heap of money and then organize your life around that. Then she tried the standard life approach with a meaning premium—she got the job supposedly working with kids. Even there, somewhere along the line Jess starts to realize she is barely staying happy enough to keep it all going.

In some ways, she is one of the lucky ones. She got a standard life approach job in line with her calling. Lots of people can't get the meaningful job. The record label won't sign them. The director won't cast them. The social work agency won't hire them. Does that mean there is no way for them to follow their calling? Should they stay in jobs they don't like while having insufficient free time to do what they want to do?

Not everyone's life is quite like that, not by a long stretch. But a lot of the lifequesters I've met began their quests for just that reason. Of their eighty-four-hour weeks, they realized they had 3,245 left and then 3,244 and then you blink and a year has passed and it's 3,192. Meanwhile, they wondered, *Why am I alive? Why was I born?*

They came to a place where they asked, *Why should I bend my entire life around how I earn my money? Why don't I bend how I earn my money around the life that is True for me, the Good Life?*

Bending Your Earning Around Your Good Life

Here is an altogether different approach that some lifequesters are taking: Instead of choosing a job that either earns a pile of money or vaguely falls in line with your calling and then fitting the rest of your life around that, choose a life in line with your calling that you enjoy and love and with sufficient leisure time, then fit your earning around that.

Let's return to Jess. Sick and tired of being sick and tired, seeing nothing on her bootstrap list really changing, she asks herself what life she would want if she didn't have to earn. She digs out her bootstrap list.

One thing she knows is that she wants to get back to surfing (15 hours a week, according to her list). Also, she wants to spend another 15 hours a week hanging out with people she loves. Plus, she wants to return to cooking and eating local food from the farmers' market. All of this tells her that it is probably time to move back to Southern California, where she grew up.

Luckily, the one good thing about her social work job is that she now knows she is called to work with kids. She feels she has a gift. She would want to do that part of the job whether she got paid or not. She wrote on her bootstrap list that she wanted to spend at least 15 hours every week working with kids. But she wanted to spend zero hours writing reports about them or meeting with colleagues.

She also knows that she doesn't want to spend time in a car commuting anymore and that she resents spending a ridiculous amount of money on work clothes she doesn't like wearing anyway. Her friend has offered her a room in a community living situation at a much lower rent than she pays now. She realizes she wouldn't have to travel across the country anymore. All of this will save her a lot of money.

Will she be able to get a job working with kids back in Southern California? She doesn't know. But looking at her bootstrap list, she sees that even if she only finds a job as a server at a restaurant earning twenty dollars an hour and working 25 hours a week, she will be able to cover her expenses and still have 60 hours for surfing, volunteering with kids, and hanging with friends.

This is called bending your earning around the life you want instead of the other way around.

Liberating Your Time and Money

What if you try to figure out what life you want exactly? Where you want to live. What kind of friends you want. Whom you want to help. How you want to live out your calling-for-now. How many hours you want to work for pay. And then try to bend that into making whatever money you need.

As a thought experiment, look at your bootstrapping list and ask yourself, *Is there a way I could reduce my expenses by 25 percent?* Maybe you could reduce your rent radically by living in community instead of living alone? Also, is there a way you could increase your effective hourly pay by 25 percent (by reducing work-related expenses or "free" time you dedicate to work or finding a higher-paying job, say?).

If so, you could have the same standard of living in a twenty-two-hour workweek as you used to have in a forty-hour workweek. That adds up to a lot more time teaching kids to sing or surfing or marching against climate change or whatever else you feel you really ought to be doing with this time of your life.

Living with a Life Satisfaction of Ten out of Ten

There are three things that are important to how thirty-three-year-old Thadeaus Umpster lives his life. One is helping his local Brooklyn, New York, community work together to become more self-reliant and less dependent on the consumerist rat race; another is living a self-reliant lifestyle himself; and a third is having a strong community of friends who share his values and with whom he gets to spend his time.

His main work—that is, the work that is most in line with his calling and that he spends the most time on—is unpaid. One branch of that work is running the Brooklyn Free Store, a pop-up store that he opens every Friday in the Bedford-Stuyvesant neighborhood of Brooklyn. People bring things they no longer need and take things they do need with no money exchanged—everything from lawnmowers to books to clothes. The other branch of his main work is publishing zines and pamphlets about social justice issues. He is also very active in the race justice and Occupy movements.

Meanwhile, to support himself, Thadeaus has found a way to earn money that enables him to live the life he wants to live. He sells secondhand books on Amazon. In 2002, he says, "I discovered that kids at all the colleges in New York were throwing out hundreds of expensive textbooks."

Thadeaus finagled a New York University ID and used it to get into dorms and leave boxes in the foyers with a sign that read DON'T THROW YOUR BOOKS AWAY. LEAVE THEM HERE AND THEY WILL GET REUSED. The boxes filled up.

At first, Thadeaus would sell the books as a sidewalk book vendor, listening to the radio and talking to people. "That was really fun, meeting people, but it wasn't worth the time." Now, he simply scans in the bar codes of the books that he gets and they automatically appear for sale on his Amazon vendor account.

These days, Thadeaus finds the most lucrative system is to collect books from law school students when the law schools let out in late July. Every year, he works a solid month until the end of August, riding his bike from campus to campus retrieving discarded books. "For the rest of the year I work barely part-time, sticking books in padded envelopes and dropping them at FedEx."

Recently, Thadeaus retrieved a suitcase full of old magazines that are selling for about $10 each on Amazon—so a trove worth a couple of thousand dollars. He has sold a single book for as much as $600. In the last year, he made $18,000 for his one month's work. It would hardly be considered a living in one of the most expensive cities in the world, but the life Thad loves living also costs him very little money.

"When I was in sixth grade I learned about homesteaders who

were living off the grid in Alaska and I wanted to do that," says Thadeaus. He was naturally attracted to a self-reliant, do-it-yourself life. "Then I realized I didn't have to do it alone in the woods. I could do the same thing as an urban homesteader." So Thadeaus moved to New York City from his hometown of Boston.

At first, he learned the ins and outs of living rent free by squatting. "But I couldn't usually stay more than six months anywhere," he says. So he instead concentrated on learning to build and live in community. He now lives with five other people and pays a monthly $400 in rent.

He also learned to scavenge. He discovered that on the streets of New York, if you go out on recycling night, you can find everything from furniture to clothes to more books. Some of this stuff he brings to the Free Store and some he uses for himself and his community.

Thadeaus has also developed relationships with a number of grocery stores. Instead of throwing food out on its sell-by date, they give it to him. He in turn cooks it together with the people he lives with. "The meal is perfectly good and I always end up living on excellent fresh, organic food."

Thadeaus has no real expenses other than his rent and his cell phone. He gets around New York by bicycle. "When I travel, I either hop freight trains or hitchhike. I have been in forty-eight states that way." Meanwhile, he has enough savings to feel safe and secure.

As for life satisfaction, Thadeaus says, "I'd say I have a ten out of ten." He has been following his calling as an urban homesteader for nearly fifteen years.

"But the reason why I am happy is because I am part of a real community of people who care about each other. Actually have time to spend with each other."

The point is not that we should all live or even want to live like Thadeaus. But he is an example of someone living his calling. He is living his dream. His way of life is what is right for him, but maybe not for some of the rest of us.

What is inspiring about his story is that having figured out what his vocation is, he has found a very unconventional way to do it. So can we. And in doing so we might get to do the many other things that make life as a human being so wonderful.

Thirteen Simple Tricks That Can Make Bending Your Earning Around Your Good Life Easier

1. **Choose your life before your job:** Live where you want to live. Hang with whom you want to hang with. Follow your calling. Figure out how many hours you want to spend doing what. Then figure out how much time you have to work. Choose your expenses accordingly.

2. **Let your real work and your paid employment be different:** Sometimes your real work and your paid employment will happen at the same time. But sometimes you can earn more and work less by taking paid employment that is not your real job. Then you get to spend more time doing your real job without pay.

3. **Don't let others define you by your paid employment:** If you find a way to live the Good Life and you are giving to the world and having a blast, the joke is on them.

4. **Drink bleach before taking on any new unsecured debt:** No credit cards. No lines of credit. Debt is the modern indentured servitude. Having to pay back old debt can mean having to stay in a job or live a life you don't want.

5. **Get help managing old debt:** One great resource is Jerrold Mundis's book *How to Get Out of Debt, Stay Out of Debt, and Live Prosperously.*

6. **Keep a tally of what you give to the world:** This is how you will be measuring the results of your efforts, as opposed to your bank balance. It will also give you the energy to go on.

7. **Build a metaphorical go bag:** Since you may not have standard means of feeling secure, develop a plan or asset base that makes you feel safe. This may be partly in the form of savings. It could also be explicit agreements with friends to bail each other out or give each other places to stay if things go south.

8. **Build a strong personal community:** This is the place where you will use your talents and skills and be acknowledged if you don't have a community through your paid employment. It is also where your security will come from.

9. **Commit to not committing:** Remember how the horse roamed until it found the tree by the stream. Don't build your corral until you find your version of the tree by the stream.

10. **Give yourself permission to try:** This is a quest. Temporary failure and U-turns are allowed. Give yourself a break.

11. **Lower your expenses:** Live in community and reduce your rent. Get rid of your car and live where you need to be. Don't buy stuff you don't need. Cook at home.

12. **Increase your earnings:** Let your paid employment pay you what you need and don't worry about doing your real work there exclusively.

13. **Participate in the gift economy:** When baking a loaf of bread, bake two and give the second away. If you do this in a community of people who share your values, you will soon find yourself in possession of some peanut butter and jelly to go with your bread.

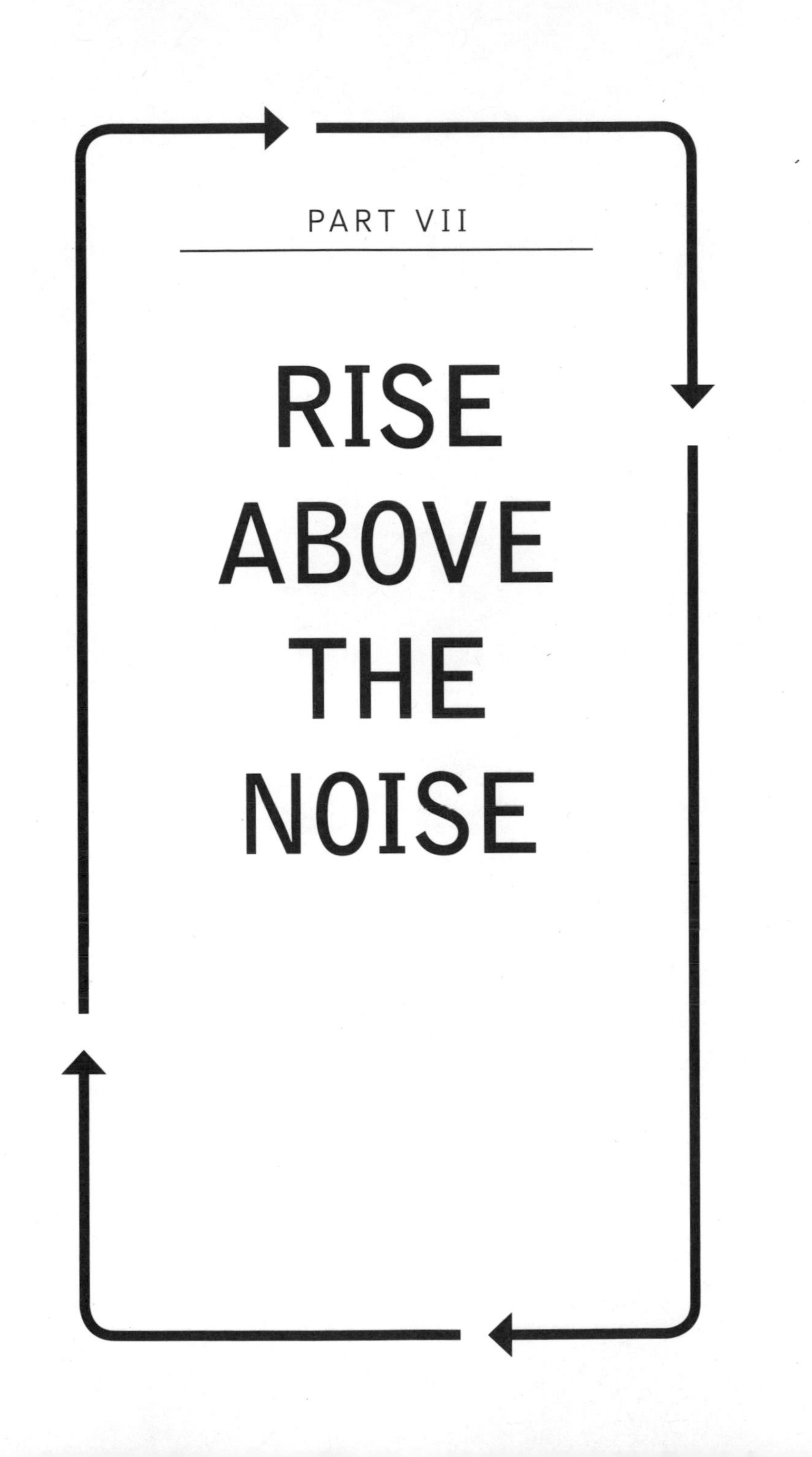

PART VII

RISE ABOVE THE NOISE

30

WHAT IT'S LIKE TO BE HUMAN

Deeper Questions That Some Lifequesters Grapple With

Without the guidance of the standard life approaches, how do I know I'm making the right decisions?

As hard as I try to do right by the world, other people don't seem to care. How do I deal with my grief, despair, and anger?

Now that I have my own version of the Good Life, I find that I am perhaps more rather than less sensitive to all of life's big questions. How do I cope with that?

Even though I am building the life I want, I still wonder, how do I deal with, well, being human?

These are some of the bigger and deeper questions that I have heard lifequesters articulate around the world. Being a lifequester, not following the societal directions, means that one must be centered in one's Self and have confidence in its guidance. But that kind of confidence comes and goes as we change as people, as the world changes, and as our lives change. Being a lifequester, for example, does not make us immune to the anxiety that comes when a big, unexpected bill arrives.

At the same time, so many of the lifequesters I have known have worked hard to help their local and planetary community. Then, say, some developer comes along and buys up all the houses in a neighborhood, essentially shattering the community a lifequester has worked to build. Or a politician approves legislation for the sake of campaign contributions, in spite of a lifequester's activism. Though we gain strength from our own and our communities' good intentions, we can't help feeling very uncomfortable feelings of grief, despair, and anger. How do we deal with them?

Not only that, but just because we are lifequesters and just because we build different sorts of lives does not mean that we face different sorts of existential questions. We still have to deal with the fact that we will all one day get sick, grow old, and die. We all have to deal with the loss of loved ones and many other of life's disappointments. Because lifequesters may embrace their humanity more fully, they may be even more sensitive to these dilemmas. How does one deal with that?

A Story About Getting Caught Up in the Noise

Samuel Miller, former dean of Harvard Divinity School, gave a lecture in which he told of a show he saw in Munich in 1931. It was a performance comprising, as Miller described it, a bunch of clowns trying to express metaphysical truth.

In it, a clown entered a circle of light on the stage and began looking very diligently for something he had lost. Before long, a policeman came along and asked, "What are you looking for?"

"The key to my house," the clown said. "If I can't find it I will have nowhere to sleep tonight."

So the policeman joined in the search. After a while he said, "Are you sure this is where you lost it?"

The clown said, "Oh, no. I lost it somewhere over there."

"Then why on earth are you looking here?"

"Because there is no light over there."

This is very funny. It points to the futility of human experience, how so often we look for things in the wrong places. We get distracted from what we know—deep inside—to be True. The clown knew where he dropped his key but he preferred to hope he could find it where the light was shining.

So many of us know what we and the world really need, but we still hope to find it in the way that everyone else is trying. Or we get distracted from our paths and calling by the shining light of our anxiety or despair or anger.

How do we deal with so many feelings and distractions and do what we know to be important and True when there are so many pressures—both internal and external—that push us away from it? How do we rise above the noise?

A Parable About Not Getting Quite So Caught Up in the Noise

Once two friends met on the road between their neighboring villages. They shook hands and asked each other how they were. One had recently had a happy encounter with the wise man from his village. He told his friend about it.

"You know," he said, "I'm sure he must be the wisest wise man in all the country."

"Well, hang on a minute," his friend replied. "I'm sure he is very wise, but let me tell you how wise the wise man from my village is."

They swapped stories, each trying to prove that his village wise man was wiser. Before long, members of both villages, walking past the arguing friends on the road, had gathered around and joined in.

"Our wise man is wiser!"

"No, our wise man is wiser!"

Things became more and more heated until finally someone said, "I know how to resolve this. Let's have a wise man competition." The crowd broke up, with one group going to each village to drag back their respective wise men.

Before we go on, hearing this story, you might say to yourself, how stupid is this? Why would anyone care about who had the wiser wise man? What a waste of time and effort! But this is just a reflection of the kind of noise we get caught up in as humans. It is an example of the noise that stops us from paying attention to what is really important in our lives.

Crazy though it might be, the wise men came, accompanied by the entirety of both villages. Even the wise men, it seems, were caught up in the noise of this ridiculous—yet entirely human—folly. The wise men stood under a tree with the crowd circled around them. The rules of the competition were that the wise men would take turns answering the other's questions until one faltered or couldn't answer.

So, each wise man took his turn and each time his competitor answered correctly. One question brought on another answer. The next question brought the next. The questions got harder and harder and yet both wise men were able to answer.

Finally, one of the wise men caught a bird from a tree branch in his hand. He put the bird and both his hands behind his back. He looked at his competitor. "If you are so wise," he said, "tell me this. When I bring the bird out from behind my back, will you find that the bird is dead or that it is alive?"

The villagers gasped. There was no way to answer the question correctly. If the answering wise man said the bird was dead, his competitor would bring it out alive. If he said it was alive, his competitor would simply twist its neck and bring it out dead.

The crowd fell quiet. The people from the answering wise man's village became angry. This wasn't right. This wasn't fair. There was no way for their wise man to win. But what does it mean to win? In a competition to be the wisest, how can you actually win? It is futile. Just as vengeance is futile. Just as envy is futile. Just as so much we get distracted by is futile. And yet we get caught up in the noise. That is just part of being human.

And it may be true that even the wisest men and women sometimes get caught up in the noise—otherwise why would the wise men of this story indulge this silly competition? But maybe, if we are really wise, we don't have to be so caught up in the noise that we can't stop ourselves from causing suffering to ourselves and those around us.

That is just what happened to the second wise man. Whether or not he realized the ridiculousness of the competition, he at least was awake to what was really important at this point. The bird's life was more important than this stupid competition. Realizing that, he knew how to answer.

"The bird is dead," he said. The rival showed him the bird—alive—which flapped its wings and flew away. That is called rising above the noise.

There is so much noise everywhere. There is the pressure to conform, to make all the standard life choices. There are the worries and anxieties of everyday life.

At the same time, if you've come to this book, chances are that you are worried and frightened by the world. Climate change is getting worse. Money in politics is out of hand. There is still huge systematic racism in our country. Each successive year is reported to be the warmest year ever recorded. And the list goes on.

How do we not get overwhelmed by fear and anxiety? How do we not decide to become end-timers? How do we not become complete hedonists, since some of us fear the end is coming anyway? How do we deal with grief over the fact that so many people don't seem to care?

How do you stay in your center and trust your process?

In short, how do we rise above the noise?

Do This!

Quieting the Storm Exercise #1

Make Calming Practices a Habitual Part of Your Day

Quieting the storm is work that takes a lifetime. But start with the exercises in this part of the book. This one was adapted from Rick Hanson's book *Buddha's Brain: The Practical Neuroscience of Happiness, Love, and Wisdom.*

Recall memories of being with someone who loves you: When something shakes you, recall the memory of being with someone who loves you and really revel in it. This releases happy-making chemicals like oxytocin into the brain. If you do this habitually, then neural pathways will be burned in and your brain will automatically begin to release oxytocin when you are upset, so you can see straight and make effective decisions about what to do next.

Inhale deeply to force a very long exhalation: In a moment of tension, when all the possibilities seem to have shrunk down to fight or flight, this activates the parasympathetic nervous system (PNS). The PNS stimulates body systems that take effect when we are not stimulated by the fight-or-flight response but are at rest. Activation of the PNS with deep, long exhalations helps us think clearly and avoid hotheaded behavior.

Shift your attention to something that makes you happy: When you find yourself reliving a situation that makes you too sad or angry to think straight, shift your attention to a memory or experience that makes you feel happy. If you do this habitually, then the uncomfortable memory will automatically trigger the happier memory, too, and the synaptic connections within your brain will actually change. This will allow you to keep your attention on what is important to you instead of what is temporarily bothering you.

Savor experiences that make you feel good: Stop and watch the sunset, enjoy playing with your dog, run in the woods, and all the while remind yourself of how lovely the experience is. Really savor the everyday moments that make you happy and tell yourself about all the details that make them lovely for you. This will increase your brain's dopamine levels, which helps you stay calm and able to focus on what is important to you.

31

THE FREEDOM OF NO ESCAPE

Delusions Are Endless

At the Chogye International Zen Center of New York, part of the Kwan Um School of Zen, where I study meditation, my fellow students and I sit cross-legged in the dharma room on two long, parallel rows of mats and cushions. It is completely quiet except that every so often, in another room, a bell rings. When it does, someone gets up and shuffles through the door. One at a time, we each get a chance to ask our private, personal questions of the Zen teacher.

On this particular day, though my body is quiet, my mind is loud. What's bothering me? The same things that bother everyone. Money problems. Kid worries. Job stresses. Relationship struggles. Nothing stays still. Everything changes. So many decisions. It just happens to be one of the periods in my emotional life when the basic insecurities weigh heavily and I am engaged in a futile struggle against them.

I am hearing, in other words, a lot of noise.

The bell rings again and it's my turn. I unfold my legs and tiptoe to the door. I slip into the interview room and perform the various standing bows and prostrations that are part of our form. The teacher gestures toward the cushion and I sit down in front of him.

"Do you have any questions?" he asks.

Questions?

Yes, I have questions. The same questions probably everyone else has. How do I make the discomfort of life go away, hopefully forever? How do I face up to the fact that I am going to die, like everyone, and stop worrying about it? How do I make it so that I don't feel the insecurity of life so keenly? How do I deal with the fact that the world is messed up and the politicians don't seem to care?

Do I have any questions?!

Almost as a joke, I say to the teacher, "Okay. Let me ask this. What should I do about my fucked-up life?"

The teacher leans forward with his hands and chin resting on his Zen stick. He smiles. He says, "Make it un-fucked-up."

Really?, I think. *That's your answer?* So I ask, "Is that working for you?"

He says, "Not so far!"

Then we laugh. Hard.

I like this. To be reminded that one of my Zen teachers can't quite get his life together. He has had his fair share of niggling money and romantic problems, I happen to know. His own life is fucked up, too. Actually, after years of trying to find someone whose life wasn't a little messy, here is what I've discovered: Even the big-time gurus are fucked up. Gandhi had a terrible temper and was mean to his wife. Martin Luther King Jr. couldn't keep his pants on.

Nowhere have I been able to find anyone who has transcended her own humanity. This is no longer bad news for me. It means there isn't something I'm doing wrong. To be reminded of the guru's inability to transcend his humanity makes me feel better about not being able to transcend my own. Maybe this is just being human.

Sitting with me in that interview room, my teacher says, "Now you know what it means that 'Delusions are endless. We vow to cut through them all.' "

He is quoting one of the Four Great Vows that guide Zen practice in my school. This one, about endless delusions and cutting through them, like the other three, can mean many different things at different times. But to me, just now, it means, "The confused view of life that comes with being human never goes away, but we vow not to get so caught up in that confusion that we can't do any good for ourselves and others."

Kind of like the wise man who deliberately lost the competition to save the bird. Yes, he got caught up in delusions at first. Otherwise, why would he have competed? He was human. But when it really counted, he didn't get so caught up in the delusion that he forgot his Truth or caused suffering for the little bird.

My Zen teacher, Gandhi, Martin Luther King Jr.—they are all human, too. So there is nothing wrong with any of us if we are having a hard time. The delusions never stop. The confusion, the desire, the anger that come with being born may never completely go away. But we can detach from them enough to do a little good.

That's why it's inspiring to know that Gandhi and King had their shortcomings. They weren't so different from us. They got caught up in the delusion, but mostly, when it was important, they were able to cut through it. They heard the noise, but mostly, when the chips were down, they rose above it.

The Difference Between Inside Fucked Up and Outside Fucked Up

This is what my teacher, during that interview, talked about next. He said, "You have to remember that there is a difference between being inside fucked up and outside fucked up."

Outside fucked up depends on circumstances and the changing emotions and feelings that come with them. The loss of a loved one. The end of a relationship. An unwanted change at work. Or even just the little things, like an unexpected big bill or the involuntary cancellation of a well-deserved vacation. We work moment by moment to respond to those circumstances, to put one foot in front of the other and make them un-fucked-up. That is natural.

Inside fucked up, on the other hand, is when you can't come to terms with the fact that you will always, to some degree, be outside fucked up. It is when you are so caught up in the mistaken idea that you can somehow stop the delusions from coming and going that you put all your efforts into barricading the doors of life. It's when you are shaken by the very process of being human.

You start to think a bigger house will fix it. Or maybe a better partner. Or a higher salary. Or maybe, if you care about social change, if everyone else in the world just starts to care about the important things the way you do. Maybe then you will feel better.

It is not that trying to get something better for yourself or for the world is wrong. But if you make the mistake of thinking these things will fix you—that they will stop the delusions from coming and going—then your hunt for them might become desperate. You might go off course—forget your calling, forget your Truth, start acting from anger and fear instead of love and compassion.

That's when we suffer and cause others to suffer. We have all done it. We shop as a fix and waste the world's resources. We stop taking time to participate in the healing of our local and world community. We stop taking time to help change the systems that cause suffering for so many. Or we do try to participate, but only in an attempt to treat our own emotional need.

We lash out at people—a taxi driver who goes the wrong way, say, or people who disagree with our vision for the world—as though somehow, if these people just did as we wanted them to, that would stop our life's discomfort. As though the process of life itself can be always pleasant. As though we can actually force something or someone to somehow make things always feel good when we can't.

Finding Real Freedom

In the interview room, what my teacher was trying to tell me was that if I can accept that the delusions—the confusions—are endless, then I will not be inside fucked up. Because then I will see that fucked up is not fucked up. It's just part of life. If delusions are endless, why get upset about the fact that we are deluded? Why be controlled by the need to get away from any one delusion? There is always another delusion to follow it.

If I can avoid being inside fucked up—if I can accept the noise—then I won't have to put my energy into trying to change what can't be changed: being human. I will have freedom.

Even though I feel hurt and anger sometimes, maybe I don't have to fight those feelings. Maybe I can instead accept some discomfort and leave myself free to invest my energy into things I really care about. I don't have to be blown around by changing life circumstances and feelings. I won't have to be distracted, like that clown, by where the light is shining. I can follow my life's True Direction. I can do what gives me long-term satisfaction instead of short-term relief.

Do This!

Quieting the Storm Exercise #2

Constantly Keep Your True Purpose in Mind

Nietzsche wrote, "He who has a *why* to live for can bear almost any *how*." The science agrees, in part because of a brain system called the anterior cingulate cortex hub (the ACC). The ACC governs what we give our attention to. When we have strong intentions, the ACC gives our attention to things that affect those intentions and ignores things that don't—like the noise of our atavistic instincts.

No matter how strong the winds of envy or desire or lust become, if your intention is strong, then you can't be blown off course. You can rise above the noise because you have something that you believe is so much more important than the noise. Remind yourself several times a day of your True purpose.

32

THE SCIENCE OF TRUTH

The Lizard in Your Brain

To some, what I have said above may sound too esoteric. Too patchouli oil and macramé. Not to dismiss the mystical approach, but if science appeals to you more, remember that brain research bears out the fact that there is often a tug-of-war going on between the "lower" and "higher" parts of the brain. Literally, one part of your brain wants you to pay attention to your outside job and another wants you to pay attention to your inside job.

The cortex, the "evolved" mammalian part of the brain, struggles to calmly make sense of the world, while the brain stem, the lizard part of the brain, fearfully puts our bodies on red alert about potential dangers or rewards that really aren't as bad or as good as it thinks. The result is that we get deluded about life and can't see straight enough to navigate through it effectively and compassionately. The brain stem can get tyrannical about whatever makes you feel safe and secure—be that a big house or better politicians in office—even if the cortex knows that right in this moment you *are* actually secure.

This is because, through the millions of years, evolution did not change or replace the animal parts of our brain but instead built more complex structures on top of them. That means we still share with monkeys and squirrels the relatively simple brain structures—the brain stem and the subcortex—that classify everything as a threat to run away from, a reward to grab quickly, or irrelevant information to ignore.

These primitive structures act quickly and intensely, often sending alarm-bell chemicals like adrenaline and cortisol coursing through our bodies. To those ancient parts of the brain, pretty much anything worth paying attention to is an emergency, a tiger that wants to kill us or a rabbit we must catch if we aren't to starve.

Meanwhile, as the lower parts of the brain start setting off alarms and sending up flares, the more complex, slower-acting cortex tries to rise above the noise while it makes a more subtle and nuanced analysis of the situation. The cortex evolved to govern more complex human behaviors like parenting, communicating, cooperating, bonding, and loving—the behaviors that make modern humans successful.

But suggestions for action that the cortex makes can have a hard time competing with the intense physical feelings caused by the lizard brain adrenaline rush. That's why heeding the call of our Higher Selves can be like trying to hear a whisper in a hurricane.

Mystically speaking, a big part of the dilemma for human beings is the delusions, angers, and fears that hijack our minds and make us believe we are separate from each other and must compete and fight to get things we supposedly "want" (even if what we want is a better world). Meanwhile, every single spiritual and religious discipline points to the fact that there is a deeper wisdom within us that naturally points us to a more compassionate and cooperative path that leads to both our individual and our collective well-being.

Scientifically speaking, that same dilemma for human beings and the source of our delusions can be explained by the fact that the social functions of the cortex are the most important in modern society, but the other structures, the ones that think we are constantly hunting or being hunted, are faster and motivate us more.

When That Lizard Tries to Run Your Life

To make matters even more complicated, the prehistoric portions of our brain also react to certain emotional states as though they were as dangerous as poisonous snakes. Because low social status, for example, used to mean you'd be the last to eat as a pack animal, the ancient parts of our brain send off alarm bells when we feel shame. Because our genes would not get passed on if we did not mate, jealousy can cause us to become aggressive. All of a sudden we are as motivated to do something about our shame or jealousy as we might be to get away from that poisonous snake.

Recently, as an example from real life, my friend's ex-wife took their son horseback riding. The son came back to my friend's house the next day and told my friend how much he enjoyed it. The son went on about how nice his mom was to take him and how he was going to horseback ride for the rest of his life. The thing is, my friend grew up horseback riding. Surely, his thinking told him, he should be the one who introduced his son to horses.

Did his son like his mom more than him? my friend found himself worrying. Did the son think his mom was more fun? What could my friend do to make the son like him better? My friend felt so terribly jealous! He found himself trying to figure out how he could take his son horseback riding, and what things he could do that would make the son say that the father was the best.

But is that what would make my friend a good dad? Trying to compete with his son's mom? Luckily, my friend realized that his concerns about the horses were crazy, useless thoughts. He was able to get back to his desire to simply be a loving dad.

The thing is, all of us have millions of crazy thoughts. The sad news is that some of them motivate us. We allow outside fucked up to become inside fucked up and then we act a little crazy. It's one thing to feel a little twinge when your daughter says she likes your ex better. It's another thing altogether when your lizard brain tries to get you to push the uncomfortable feeling away by scheming to one-up your ex.

All this lizard-brain stuff might be okay if you felt your entire purpose was about finding food, avoiding snakes, and protecting your pride. But if we have some more subtle purposes—like loving each other, helping each other, walking our individual life paths—the perception of life as merely a series of short-term rewards and threats gets in our way.

Our animal instincts, our fears, and our angers can overwhelm us. They can make the quiet voice of our True Selves difficult to hear, trust, and follow. For those of us at the beginning of our lifequest, they make it hard to believe that we will be okay if we break away from the standard life approaches.

For the experienced lifequesters who feel secure in their intentions to help the world, fears and angers can be even more insidious. We might have the erroneous thought, for example, *My intentions are to be helpful. This person is frustrating those intentions. My angry actions toward that person are all the more justified by my helpful intentions.*

Thus, the same sorts of delusions that motivate a materialistic person can manifest as self-righteousness in the lifequester. If we are not careful, the angry actions hidden behind our intention to create a better world can actually perpetuate the exact problems we are hoping to change.

The Evolutionary Potholes We Get Stuck In

It turns out that there are certain types of crazy thoughts that evolution makes our minds loop through. For example, the hardwired and formerly adaptive instinct to hoard resources in the austere circumstances of the ancient world appears, in the modern world, as greed or envy.

We can get caught up in the desire for more and loop through strategies and feel uncomfortable when we see someone has something that we don't have, when, in fact, we already have more than enough. In a prehistoric world, again, it might be natural and necessary to scheme to gain possession of the sparse food and shelter that someone else controls, but that instinct, when it emerges in modern times, is called envy.

It is not to say that greed or envy are bad or good. But we can get overly attached to them. They are ruled by the powerful and intense lower parts of our brain and we can find ourselves in psychological loops, motivated to "get" or "win" when getting or winning more than we already have won't help advance us on our life paths.

These psychological loops are like emotional potholes that we get stuck in. These potholes are challenges for everyone, but for the lifequester they can bring up feelings of doubt. Though we may have chosen a path that includes less ownership of stuff, when a friend or family member drives up in a fancy car, twinges of envy are still natural. They are part of our humanity. It is helpful to know that they may well be just the noise of our vestigial lizard brains.

In addition to envy and greed, there are a number of other psychological potholes we can get stuck in, including the desire for revenge; the need for rest; our social status; our sex and relationship drives; and our relationship to addictive substances, from food to drugs to alcohol. These behaviors are all governed by the lower parts of our brain, instinctually based, and related to survival in the primitive world.

When these motivations are frustrated, our brain stem, with its arsenal of chemicals and neurotransmitters, can tyrannize our bodies into feeling that our lives are threatened when, in fact, they aren't. When someone fails to respond to a message you sent on an online dating site, your brain can react with as much sadness as if the only available mate in your nomadic tribe had chosen someone else instead of you. When someone fails to use the recycling bin when you've asked him to, your brain can react with as much pride as if you had been slighted and lost status in your group.

That is, of course, delusion. Our brain stems can force us to act as though our lives are screwed up when, in fact, they are not. We start to mobilize enormous personal resources to change things that either can't be changed or aren't worth the effort. We bring cannons to small arguments and bulldozers to minor frustrations. This can cause suffering for ourselves and others, take us away from our True Purpose, and cause us to act in ways that aren't in line with that purpose.

And, Still, Delusions Are Endless

The bad news is that we can't stop the delusions any more than we can remove our brain stems. The good news is that we still have the cortex with all its subtle functions and ability to analyze situations from a higher perspective. It can't stop the messages that the brain stem sends us, but it can help us to not be ruled by them and to greet them with equanimity.

Equanimity means we neither struggle excessively to fulfill our desires and avoid our aversions nor to suppress those desires and aversions. It means letting them come and go without overreacting to them. In other words, we let our control rest in the cortex, in the language of brain science, or in the Higher Self, in the language of spirituality and religion.

We don't have to chase after desires that divert us from our True Path. Nor do we have to be ruled by a fear or despair of the world that stops us from helping it. We cannot stop the noise, but we can rise above it. We can let the course of life be steered by our abiding wisdom instead of being at the mercy of constantly changing psychological discomforts.

To do that, it turns out that we can actually rewire our brains for more equanimity. Each time a thought passes through our brains, it causes connections to form between neurons. The thoughts that flow through our brain actually sculpt it, like water carving a ravine in a hillside. So we can use certain mental exercises—as well as spiritual practice and personal growth—to change our brains and help us rise above the noise.

Do This!

Quieting the Storm Exercise #3

Develop a Meditation or Contemplative Prayer Practice

Studies show that meditation literally grows and strengthens the higher parts of the brain. It causes an increase of gray matter to grow in the insula, hippocampus, and prefrontal cortex. These are the slow-acting parts of the brain that govern our higher social functions, and strengthening them helps them win the tug-of-war with the fight-or-flight brain stem.

Meditation also stimulates and strengthens the parasympathetic nervous system (the PNS, mentioned in chapter 30). All of this means that meditators have more equanimity, less stress, and more concentration. In short, meditators have a greater ability to rise above the noise and pursue their True Purpose.

Meditation and the religion of your choice: Meditation is often associated with Buddhism, but every religion has its contemplative traditions. You can get meditation or contemplative prayer instruction in the Christian, Jewish, Buddhist, Islamic, and Hindu traditions.

Regular practice is most important: You may already have your own favorite method, but you need to get to it persistently. Starting to meditate is a little like starting to run or exercise. At first, just set aside ten minutes each day. Don't try to do an hour. It will grow by itself.

Join or form a community of meditators: This is probably the best way to keep your meditation practice regular.

Here's a simple, secular approach to meditation:

- Commit to a time period of ten minutes or more. Tell yourself you'll meditate till the time is up.
- Find a place to sit quietly in a chair or on a cushion where you won't be disturbed. What position you are in is not as important as being able to keep still in that position. A quiet body makes a quiet mind.
- Once you are still, with your eyes open or closed, bring your awareness to the sensations of breathing, without trying to control it. Just watch.
- When your mind wanders, just bring it back to paying attention to your breath. While paying attention to the breath, let physical sensations and mental formations rise, take shape, stabilize, disintegrate, and disappear.
- Because your mind was built to think, it will tend to attach to some of the sensations and thoughts that arise and suddenly you will realize that you have been carried away. This is not failure. It is part of meditating, and becoming aware of this process helps build your understanding of what the human mind is like. Don't interfere or judge the process. Just watch it.
- When you realize you've drifted away, return your awareness to your breath. Over and over.

Peace in Yourself Helps Bring Peace to the World

All this talk about equanimity and how to cut through our delusions might seem self-centered, but the ripples spread. It's said that a peaceful mind makes a peaceful person makes a peaceful family makes a peaceful village. The opposite is also true. When we allow outside fucked up to become inside fucked up, we act a little crazy and that makes the world around us crazy, too.

Our fight-or-flight reflex makes us shout at the taxi driver, which triggers his fight-or-flight reflex. He then shouts at three more customers, who in turn shout at three more people, and the suffering spreads. The Indian philosopher Jiddu Krishnamurti said the problem with the world is our relationship with ourselves, which then becomes our relationship with one another, which then becomes the whole interconnection of relationships of the world.

So what might happen to the world if we all worked more wholeheartedly on our relationships with our Selves? If we took the time to understand our place in the world from a spiritual and metaphysical perspective? After all, this is the approach taken by some of the world's most effective change makers: from Mahatma Gandhi to Martin Luther King Jr. to Nelson Mandela to the Dalai Lama to Archbishop Desmond Tutu.

In no way am I suggesting that our personal spirituality can replace an active engagement with the suffering of the worlds we live in. Sitting in meditation or kneeling in prayer has never been intended to help us escape from the Truth of living in a material world. Instead, they are two tools, among many, that humanity has developed to help itself gain sufficient strength and peace to face the problems of our own lives and the lives of others and to constructively engage with them with joy, compassion, and effectiveness.

In spirituality, as in so much else in this book, the great paradox and life's most wonderful joy is that when we do what is necessary to care for ourselves, it will help the world, and when we do what is necessary to help the world, it will help us. That is Truth if you choose the path of the lifequester.

And I hope you will.

Epilogue

THE LAST WORD

There is a story in an ancient Zen Buddhist collection known as the *Wu-men Kuan* about an old and enlightened Zen Master, Te-shan, who came trundling down from his room in his monastery with his eating bowls even though it was not time to eat.

One of the Zen Master's senior students, whose name was Hsueh-feng, saw the Zen Master carrying his bowls and said, "Old Master, the bell has not yet been rung and the drum has not yet been struck; where are you going carrying your bowls?"

At that moment, Zen Master Te-shan turned around and, without a word of reply to Hsueh-feng, went back to his room.

In thinking about this story you need to know that part of the way Zen students of the time learned from their teachers was to ask them questions or make statements to challenge them. The teachers would then "hit" their students with a reply—in either words or actions—that pointed to some error in the assumptions the student made in the statement or question.

On this occasion, Hsueh-feng took the Zen Master's silently turning around to be a figurative answer to his question. He didn't understand this answer, so he told the story to Te-shan's deputy in the temple, the head monk, whose name was Yen-t'ou. Yen-t'ou listened and then said, "The poor old master does not understand the last word of Zen!" In Zen stories, "the last word of Zen" is taken to mean the most important teaching of Zen.

As some commentaries on this story go, Hsueh-feng may have interpreted this answer as "The poor old master does not understand the last word *as well as you do.*" Hsueh-feng thought, *What Yen-t'ou is telling me is that I have finally challenged the Zen Master in a way that proves that, at least in this moment, I had a better understanding of Truth.* Hsueh-feng's pride swelled a little. *I'm finally beginning to get some real understanding of Truth! I got one over on the Zen Master!*

Then, suddenly, according to the commentary, he may have asked himself, *But how did I do it? What did I say that was so smart? And what does Yen-t'ou even mean by "the last word of Zen"?* Suddenly, after just moments ago feeling he must have finally gained some understanding, now he felt as lost and as confused as ever. *I'll never understand life,* he may have thought. *And I don't know the last word, either.*

To say *I don't know the "last word"* is kind of like saying *I don't know how to live, since I don't understand the most important instruction Zen has to offer.*

That may be a little what it feels like to finish reading a book called *How to Be Alive* (certainly, it is what is like when you have finished writing one). For a while, you might be all excited by a bunch of new ideas and have this feeling that you finally understand something. Then you realize that life is still life and the mystery is still the mystery and, maybe a little bit, the world is still fucked up.

What are you supposed to do about all that? So, you turn to the last part of the book, titled "The Last Word."

If we are talking about how to be alive, after all that we have said, what is the last word? What do we do now? What is the capper? What is the thing that will give us that understanding? What will make everything better? Can we help in any meaningful way? What is the last word of Zen? Or, for that matter, Christianity? Or any of the religions? Or psychology? What is the last word that will make us understand it all? What is the last word that will tell us what to do?

Don't worry. All of us are confused (even those of us who think we aren't, who actually are the most confused). That is okay. None of us know. That is okay, too. Not knowing is good. Not knowing lets the Truth come out.

In Zen, the question, after all the reading and all the talk, not knowing what to do and holding on to the question "What should I do?" *is* the last word. All the learning, all the ideas, all these thoughts, all the understandings, all the new nonstandard approaches of doing things we have discussed are still just ideas that can stand in the way of our True, compassionate Selves emerging in every moment.

So the last word in Zen, after all that we have said, is "In this world, in this situation, in this life that I have right now in this moment, what should I do?" Or better yet: "How can I help?" The point is that holding on to the question and the intention moment by moment is better than all the concepts and rigid ideas we can come up with. So much of what we have said in this book is about trusting that intentionality.

There is another Zen saying that, in many ways, expresses the last word on life:

When a hungry man comes, feed him.

When a thirsty woman comes, give her something to drink.

In Judaism and Christianity, the question about how to be alive, the last word on life, is just the same:

Love your neighbor as yourself.

The thing is, in antiquity, when these expressions were coined, humanity largely lived in villages or neighborhoods. Feeding the hungry man who came, giving water to the thirsty woman, loving your neighbor as yourself may have actually solved all the problems.

Today, though they are still the right things to do, it is hard to see how these strategies can be the "last word" when the situation, say, is that oil companies are about to start drilling in the Arctic. What is the last word when we are up against huge, complex planetary problems instead of just hungry and thirsty neighbors? What should you do? How can you help?

When a hungry man comes, feed him, yes, but what about when an international corporation comes that continues to push a revenue stream that is destroying the habitat? To be effective, you have to really work to understand the complexities of the world we live in and the systems we lumber under. How can we change these big systems that seem beyond our power to change?

Right here, right now, what is the last word?

Some months ago, I was on a conference call with a number of very senior activists in the climate movement. Among them was Van Jones, the CNN commentator and former White House green jobs czar. Also on the call was Gus Speth, the environmental lawyer and advocate who cofounded the Natural Resources Defense Council.

During the call, Speth asked Jones, "Van, you've worked closely with the current presidential administration. What will make this president think that doing something about climate change will be his legacy? What do we need to say?"

"I don't think anyone knows," Jones answered. "What I would say is that if you get a chance to speak to people of influence, *you should use whatever argument you are most passionate about. Because it may not be what they can hear best, but it is what you can say best.*"

Remember this Buechner quote: "The place God calls you to is the place where your deep gladness and the world's deep hunger meet." What is most True for you is how you can most truly help the world, and what helps the world is ultimately what is most True for you. Investigate. Follow Truth. Even if it is complex and you can't see around its twisted corners.

Finding ways to help the world that make you happiest will ultimately make the world happiest. That is an article of faith in this book—letting your passion be your guide. What works for you? We have talked about everything from changing your relationships to stuff and food and transportation to becoming an activist to building social interconnectedness to sharing children to rising above the world's noise through meditation or contemplative prayer.

Becoming who you are will give you energy, but it is a process that is never over. Asking what is the last word is a process that will never end. In each moment, how can I help? Not just for me. Not just for the world, either. For us. What is the last word?

But will all this personal transformation and helping and joining work? Is there hope for humanity? Will we make it?

I was asked to contribute to a book called *Global Chorus: 365 Voices on the Future of the Planet*. It is a lovely book with entries from everyone from the Dalai Lama to Desmond Tutu to Maya Angelou. The question each contributor was asked to answer was: "Is there hope?" My entry reads:

> *The Buddha always refused to speak about unanswerable questions like whether there is life after death. He said trying to answer such questions is like a man who has been shot with a poison arrow who, instead of removing the arrow, insists on finding out who shot the arrow, who made it and how long it is. The most profound question in life, the Buddha believed, is not What happens when I die? but, just in this moment, How shall I live? What is my function?*
>
> *Asking if there is hope for humanity is a little like asking one of Buddha's unanswerable questions. If you are walking down the sidewalk and a car runs over a child, what do you do? Do you stop and ask, is there hope for this child? The question itself distracts you from doing what is important. You must run and help the child.*
>
> *Now, at this moment in history, the world is like the child. You must help it! Me too. We all must. Don't get distracted!*

This sounds like a terrible responsibility, but it is actually a wonderful opportunity. The world has so many problems that it needs all our special skills and talents. It needs scientists and economists and singers and musicians and children and Christians and Hindus and people of every type. What makes this such a wonderful moment is that, with so much trouble, each of us can make a difference.

So don't waste time asking if there is hope for this world. Rather, what can you do right now—as the amazingly special and uniquely talented person you are—to pull out the poison arrow, to save the child on the sidewalk, to help this suffering world? The question is not whether the world has hope. The question is, how do I give this world hope? Or more simply, how can I help?

How can I help?
The last word.

Resources

How to Join the Ongoing Conversation and Get the *How to Be Alive* Workbook

This book is just the beginning. You can get even more support in building a life that is good for you and good for the world. You can also *give* support to others who are trying to build the Good Life by joining the conversation and supporting the work. I hope to talk to you soon!

Get the Workbook

As a bonus to help you on your lifequest, I have created the *How to Be Alive* Workbook. The workbook is available free for download at www.ColinBeavan.com/HowtoBeAliveResources.

Inside the workbook you will find:

- Worksheets to help you with the exercises in this book
- Additional exercises
- Links to webpages that will help you achieve your lifequest goals, arranged according to the chapters of this book

Join the Conversation

Obviously, this book is not the end of the conversation about the quest for happy lives that help the world. There is more to say and explore and think about. To get the *How to Be Alive* newsletter and hear about blog posts, podcasts, and other resources that can help you and others on your quest, go to www.ColinBeavan.com/join.

Support the Work

Expanding the conversation about how to be alive in a way that fixes our world problems is crucial. Our entire society needs to revamp itself in a way that facilitates a happy, safe, fair way of life that keeps our habitat healthy. There are many ways to support my continuing contributions to this conversation—from volunteering to offering financial help. Please go to www.ColinBeavan.com/Support.

Acknowledgments

We think we cannot be heroes because we look at Martin Luther King Jr. or Mahatma Gandhi or the Dalai Lama or Bill McKibben or Bernie Sanders or whoever and think "I can never be like them." The thing is, more of them is not all we need.

We also need more of us. The everyday heroes. The people who dare to be themselves even without the acclaim, support, and acknowledgment that these well-known *super*heroes get. If we all lived in line with the real Truth of our deepest values and insisted that the institutions and organizations that we are part of also adhered to those values, the work would be close to done.

There are so many people who are doing that—living in line with their values and being happier and helping their local and planetary communities as a result. Everyone from the community gardeners to the mentors of children to passionate social workers and teachers to the freegans to the organizers of marches and rallies to the quiet friends who sincerely just do their best.

These are the people I want to acknowledge first. Since I published my previous book, *No Impact Man*, I have been in touch with so, so many people with the most generous and compassionate hearts. These unrecognized heroes inspire me. People who trust their love and act on it rather than on the fear and anger that we all sometimes feel. Thank you. Thank you. Thank you.

Some of you have had your stories included in this book. Particularly I would like to thank Jay Coen Gilbert, Rabbi Steve Greenberg, Kate Zidar, Rebekah Schiller, Monica Hunken, Thadeaus Umpster, and Wendy Peterman. Also, thank you to the many Facebook fans and friends and blog readers who have encouraged me and answered my questions and challenged me to think—many of you, too, have contributed your stories to this book.

The biggest struggle for me in writing is not the research or the synthesis of ideas. The biggest struggle for me is that my demons live at my desk. They wait for me there. When I sit in front of my computer, they tell me terrible things about myself and my work and try to stop me and torture me. Then, my poor friends and family suffer through my needing reassurance that all the things the demons say are not true.

That is why the people who love and support me have always come first, not last, in the acknowledgments of my books. Love and gratitude to: the amazing artist and mentor Maxwell Gimblett; my fellow traveler in daddyhood and being a single man again, Ryan Harbage; she who really saw and loved me when I most needed to be seen and loved, Joanna Bock; my brothers, Tanner Freeman, Sean Sakamoto, and Gary Rabinowitz; my mother, Judy Beavan, and my father, Keith Beavan, and his wife, Beth Beavan, and my sister, Susan Oliver; and Helga Grunberg (Yes, I just mentioned my therapist!).

Michelle Conlin: Thank you for being the best co-parent and also for believing so fully in me as a co-parent. Thank you also for reading this manuscript and being chief cheerleader. Whatever we have been through and whatever we will go through, you and I continue to give each other the best gift either of has ever been given—Isabella. We will always be family.

Maryanne Murray, thank you for your support and advice with the diagrams.

Thank you to Lilly Belanger, who for four years now has run the No Impact Project on my behalf and shepherded many tens of thousands of people through No Impact Week. You have been my stalwart professional support and my good friend. Thank you also to Lindsay Buchanan and Stephanie Bleyer, who ran No Impact Project before Lilly.

Since I first began the *No Impact Man* project, I have made many collegial friends, some of whom have been professionally investigating the Good Life far longer than me. Thank you for helping and supporting me in your ways: Julie Schor, Annie Leonard, Bill McKibben, Steffen and Rachel Schneider, and Julia Butterfly Hill. Professor Tim Kasser, you have been so generous in sharing your thoughts, work, research and time with me.

I'd like to thank my teachers in the Kwan Um School of Zen, most particularly Zen Master Wu Kwang (Richard Shrobe), Zen Master Soeng Hyang (Bobby Rhodes), Paul Majchrzyk JDPSN, Ken Kessel JDPSN, and Dr. Steve Cohen JDPSN.

In the New York literary community, I am so lucky to have found colleagues who are so loyal to my work and to me personally. Eric Simonoff of WME/IMG has been my agent since 1999. Denise Oswald of Dey Street not only made this book so much better but also *No Impact Man*, back when she was at Farrar, Straus and Giroux.

Thanks to the amazing editorial, production, marketing, and publicity teams at Dey Street including: Trish Daly, Laura Cherkas, David Palmer, Joseph Papa, Michelle Podberezniak, Michael Barrs, and Tanya Leet. Big thanks to Julia Gang and Michael Accordino for the amazing cover and Paula Szafranski for the page design.

Finally, in memory of Sarah Douglas. You were my oldest friend, and back when we were young, you were the one who taught me to be aware of and take responsibility for the world. The three children you left behind are becoming such lovely young men and my admiration for Michael continues to grow as he remains the adoring and loving father he always was.

About the Author

AYLA CHRISTMAN

Writer and social change activist **COLIN BEAVAN** attracted international attention for his year-long lifestyle-redesign project and the wildly popular book, *No Impact Man*, and the Sundance-selected documentary film that it inspired. He has appeared on *Nightline*, *Good Morning America*, *The Colbert Report*, *The Montel Williams Show*, and NPR, and his story has been featured in news outlets from *Time* magazine to the *New York Times*. A sought-after speaker by wide-ranging audiences, he also consults with businesses on improving eco-friendly and human-centered practices. Beavan is the author of two other books, *Operation Jedburgh: D-Day and America's First Shadow War* and *Fingerprints: The Origins of Crime Detection and the Murder Case That Launched Forensic Science*, and his writing has appeared in *Esquire,* the *Atlantic*, and the *New York Times*. He is the founder of the No Impact Project and a dharma teacher in the Kwan Um School of Zen. He lives in Brooklyn, New York.